So You Want to Build a House

So You Want to Build a House

Peter Hotton

Drawings by Gino Carpinteri

Little, Brown and Company

Boston — Toronto

D

T 04/76

LIBRARY OF CONGRESS CATALOGING IN PUBLICATION DATA

Hotton, Peter.
So you want to build a house.

Bibliography: p.
1. House construction — Amateurs' manuals.
I. Title.
TH4815.H67 690.8 75-38559
ISBN 0-316-37386-9

Designed by Janis Capone

Published simultaneously in Canada
by Little, Brown & Company (Canada) Limited

PRINTED IN THE UNITED STATES OF AMERICA

To my wife, Lucia,
who has spent
half her married life
in sawdust

Acknowledgments

I am deeply indebted to:

Louis Bonaiuto, my father-in-law, who helped me in many of my projects, who taught me the shortcuts, who good-naturedly reprimanded me when I made hammer marks in perfectly good pine woodwork and showed me how to prevent such butchery, and who proofread parts of this book.

Richard Connolly, whose brain I picked for a series of articles in the *Boston Globe* that showed, step by step, how he subcontracted the construction of his own house.

The dozens of individuals who opened their houses to me for articles in the *Globe;* houses, old and new, they had spent long hours in building and making livable.

The hundreds of individuals, who, when they call and write me for advice, make me do my homework in keeping up on new products and techniques.

The manufacturers and manufacturing associations, who send me all kinds of information, some highly commercial, some with good ideas and good advice, all of which I store away for reference.

The lumber dealers, roofers, concrete form men, excavators, and all the rest of the men in the building trades whom I have called when I was stuck for an answer, and who helped me on projects by doing their work and talking enthusiastically about it.

Finally, my father, Nicholas, who let me build a shack in the backyard many years ago back home in Michigan, and who spent Halloweens guarding the place so it wouldn't get wrecked.

Contents

Introduction

There are too many bad jokes about inept householders painting themselves into a corner, or bloody-thumbed handymen struggling halfway through their projects only to end up having them finished by a pro or, worse yet, their wives. Such stories have some truth in them, but there are many more men and women who have successfully renovated old houses, built additions, decorated, and done all the things necessary to make a house a home.

So, what is needed to build a house from the ground up?

Experience helps; so do good instructions; even better is a knowledgeable friend or relative. But to build a house, one also needs confidence in one's own ability, a practical sense of timing, plenty of patience, and above all, a passion for working with one's hands.

This book tries to make the builder's chore a little easier. Experienced do-it-yourselfers may find it too elementary (they know the sizes of lumber and nails, for instance); novices might find the chore overwhelming. No one in his right mind says building a house is easy; pitfalls are pointed out; different materials and techniques mentioned may be confusing; but if nothing else, it should be fun.

So You Want to Build a House

chapter 1

Don't Be Bashful, Don't Be Scared

Three Ways to Build a House

Building a house is a challenge to anyone who is planning his dream house. If he doesn't think it will be, he'd be better off going out and buying one already made. Actually, there are three ways to "build" your own house:

1. Find some contractors, invite bids on the type of house you want, and let the lowest (or best) bidder worry about construction, weather, insurance, financing, scheduling, buying materials, and hiring subcontractors. Then you can visit the site once a week or even once a day to make sure your desires are being met; but let the builder do the worrying.
2. Act as your own general contractor; buy the land, get permits and financing and invite bids for materials and subcontractors to do various parts of the house.
3. Act as your own contractor and do as much of the work as you have time and talent for. You may feel you can do different kinds of work, such as framing, laying floors, putting on sheathing, siding, and roof and roofing shingles, installing windows, installing insulation, putting up dry wall or plaster, setting tile, and doing finish work.

The second and third methods are interchangeable; you simply decide what you can do and do it. If you have owned a house in the past, you probably have done some of these things. Your attitude might be great but your aptitude might not; the important thing is to find out if you can do various jobs well and enjoy them. Nothing is more frustrating than discovering that you want to chuck a project once you're committed.

If you find you're talented in some aspects of building a house, you're halfway home. But a word of caution: laymen are generally very slow in their work, no matter how eager they may be. They will probably take two to four times what a professional will take, man for man, hour for hour. One of the reasons for this is that the work being done is just twelve or so inches away from the worker's nose; he sees all kinds of defects close up, and worries and fusses and takes too much time trying to fix them even when they don't need fixing. When the house or project is finished, the "defect" is either hidden by other construction or is far enough away from the viewer so that it's invisible.

Some builders and professional people in the field will discourage a layman from doing too much on a project. On the other hand, some professionals are extremely helpful and accommodating, especially if they are hired to do some of the work.

This book will describe the second and third ways to build a house, so the prospective builder-owner can make up his own mind. Some do-it-yourself books disparage the pros, saying that the layman can do virtually anything. Well, we've met too many professionals to knock them. They are hardworking, honest for the most part, and, man, are they ever knowledgeable. And considering that there are dozens of trades involved, that's a lot of knowledge. If a pro is willing to part with some of that knowledge, you have a friend.

Even if the builder-owner doesn't do all the things described here, it will pay him to have a working knowledge of them so he can talk intelligently, but not smart-alecky, to subcontractors.

Perhaps the first thing to study is the glossary of terms at the end of the book. A working knowledge of the terms can give you equal footing with the pros, more or less, when inviting bids and discussing problems.

The phases of building from below ground to roof can apply to building an addition to your present house. That might be just the ticket, because with costs what they are today, it may be easier and cheaper to move up (a dormer or second floor), out (a wing), or down (in the basement) than moving away.

How much can you expect to save by contracting your own house? In dollars, that's tough to say, because of the size of the project and such things as cost of land. But in percentages, 10 to 20 percent is not unreasonable if you do a minimum of the work· up to perhaps 40 percent if you do much of the work yourself. To give just an idea of what tradesmen get these days, masons get about $75 a day; carpenters do about as well, and there are few tradesmen that you will hire who earn less.

How many of the phases of work can you expect to do, given adequate ability, time, stamina, and enjoyment? Let's put it this way. What can't you do, either because of heavy equipment or specialized, expensive tools involved in a job, or because of strict codes requiring the work of a licensed man?

For instance: Excavation. Unless you have a brother-in-law who has a backhoe and/or bulldozer, this is out of the question. Besides, it really isn't that expensive to dig a hole properly.

Pouring concrete footings and foundation. Footings must be level both length- and cross-wise. You can build forms for footings, but it's exacting work. We know of one layman who built a form for concrete out of plasterboard, and was insulted when the concrete man said it wouldn't hold the concrete without buckling.

Foundation. Even tougher, because you might be able to rent foundation forms, but what you'd take a week or a month to do, a skilled crew can put up in a couple of days. If you put up concrete blocks instead of poured concrete, the time involved would be discouraging. However, Winston Churchill laid bricks to soothe his nerves during the London blitz, so it amounts to different strokes for different folks.

Electricity and plumbing. These are closely watched by community officials, and permits are needed and inspections are rigid. If you have a relative or friend who is an electrician or plumber, you might be able to hire yourself out as a helper, working under his supervision, if the code allows it. You'd do the heavy "bull work" and the pro would make the proper connections and other critical phases.

Heating. A heating plumber must work on hot-water heat. With hot-air ducts (sheet metal) you could do the work, but it is best done with supervision in order to balance the system and make sure the registers are in the right place (under the windows is best, not on an interior wall, and low, not high), as well as the cold air returns.

Sewer or septic tank installation. The former must be done by the community. No ifs, ands, or buts. A septic tank installation may be a do-it-yourself job, but it requires digging, and we're right back to the requirement of equipment.

These seem like a lot of things you cannot do, or shouldn't, but it does leave plenty: framing, sheathing, dry wall, paneling, interior trim, installing kitchen cabinets, laying tile, painting, stain both in and out, and siding and roofing.

And, there is plenty to do that requires legwork, patience, telephoning, and common sense before the first spade of earth is turned.

First, hire a lawyer. That's something else you can't do yourself, and it shouldn't cost too much. Any hassle you might get into, just turn it over to the lawyer. He can represent you in financing, closing deals, and signing contracts. And once you've found a suitable lot, go down to the town or city building department and buy a building code book, which might also include zoning regulations.

The organized person is going to keep track of everything he does, and every cent he spends on his project. If you're not organized, this is doubly important, so keep papers, permits, and bills in folders. A briefcase is a good investment for toting those important papers around. Besides bills and permits, you should make up and keep up two important papers. One is a schedule of jobs, the order in which they must be done so that the house progresses in good order and workmen don't get in each other's way. The other is a list of jobs and materials, date of payment, and check number. Make copies.

Even if a lot of jobs are done by you, you still should order and schedule them, so the order of construction can be followed to prevent chaos. And if you do a job and are not paid for it, that job still should be included in the payment schedule, to help you determine just how much you have saved by doing it yourself.

JOB SCHEDULE

Job	*Ordered*	*Completed*
Land Survey		
Excavating		
Footings and foundation		
Steel beams and columns		

Job	*Ordered*	*Completed*
Rough carpentry		
Sills		
Framing		
Sheathing		
Windows		
Roofing		
Siding		
Exterior paint		
Rough plumbing		
Rough electric		
Rough heating		
Chimney and fireplace		
Insulation		
Dry wall or plaster		
Floor topping		
Finished hardwood		
Underlayment		
Kitchen and bath		
Appliances, cabinets		
Heat registers, baseboards		
Furnace		
Finish carpentry		
Doors and frames		
Baseboards		
Stairway		
Finish plumbing		
Finish electric		
Sewer connection		
Septic system		
Water		
Well		
Electrical connection		

Job	*Ordered*	*Completed*
Bath, kitchen tile		
Interior finish		
Paint		
Wallpaper		
Stain		
Varnish		
Hardwood floor sanding, finish		
Linoleum, floor tiles		
Carpeting		
Exterior steps, porch		
Caulking		
Exterior painting, staining		
Grading		
Grass (best in autumn)		
Landscaping		

PAYMENT SCHEDULE

Job or Material	*Paid*	*Date*	*Check No.*
Appliances			
Architect			
Attorney			
Blueprints			
Bricks			
Cabinets			
Caulking			
Carpentry			
Rough			
Finish			
Concrete			

Job or Material	*Paid*	*Date*	*Check No.*
Deeds			
Doors			
Dry wall			
Electric			
Rough			
Finish			
Temporary			
Excavation			
Flooring			
Floors			
Gas connection			
Grading			
Heating			
Insulation			
Insurance			
Insurance claims			
Linoleum			
Lot			
Lumber			
Rough			
Finish			
Mortgage application			
Painting			
Permits			
Construction			
Electric			
Heating			
Plumbing			
Water			
Well			
Plastering			

Job or Material	*Paid*	*Date*	*Check No.*
Plumbing			
Rough			
Finish			
Roofing			
Sand and gravel			
Septic system			
Sewer connection			
Siding			
Steel beams and columns			
Storm windows			
Survey			
Taxes			
Tiles			
Water connection			
Well			
Windows			
Miscellaneous			

It's a wonder houses aren't built with red tape instead of wood. But if you follow the codes and have competent, honest people both on the jobs and in the community offices, you'll have a minimum of red tape.

There are permits, generally a construction permit that requires three inspections by the community building department: after the foundation is installed but before backfilling; on framework and sheathing, before insulation is installed; and a final inspection that, if you pass, allows you to get an occupancy permit. Then there are the specialty permits and inspections: plumbing, electricity, gas, public works (water and sewer connections), and heating.

Just how long should a house take to build?

With a minimum of doing it yourself, the gestation period is about nine months. That is, from turning over the first spade of earth to occupancy. Acquiring land, surveying it, getting permits, and financing might take a little longer. But if you look for land, get financing, and do other little chores during the winter, and if you start actual construction in early spring, you'll be having Thanksgiving dinner in your dining room, prepared, of course, in the dream kitchen of your dream house.

chapter 2

First Things First

Land, Financing, Improvements, and Insurance

There is plenty of work to do when building a house, even before the first earth moving starts. You need financing, land, surveys of the land to permit a septic system (if there are no sewers), a well (if there is no water service), a check of zoning laws, plans of the house and its location on the land in order to get the required permits, and insurance. This sounds like trying to surmount the impossible, but it is not. The rules and regulations, while numerous, are relatively simple, and most people are just not familiar with them.

So get someone who is familiar, a lawyer. He could be a friend or acquaintance, so much the better. He should practice in or near the community where you are building. Lawyers don't come cheap, but they can save you a lot of legwork and aggravation that might discourage you before you even get started.

How to finance everything? If you already own a house, and plan to sell it before you build or sometime during construction, your only problem is to agree on the time to vacate it. As a current owner you shouldn't have too much difficulty in getting a loan. Unless you can pay cash for the property, or can pay for it from the sale of your present property, any mortgage or loan you take out should include the land, plus improvements, and the cost of building.

One advantage in using a general contractor (rather than acting as your own) is that he can get a loan more quickly and easily than you might be able to, at least for the construction. If he's in the business of building, and his credit is good, that's why.

First, shop around. This will be necessary to obtain the best interest rate, and to find out just who will give you a loan. Not that your credit is bad, but when money is tight, lenders tend to be choosy. Your lawyer could help you with this.

Mortgage applications are fairly standard. They cover routine information, vital statistics covering you and your family, employment, assets, liabilities, bank balances, past performances on loan payments dating back two to five years, and your general credit. It sounds like the Inquisition, but if you're clean, and the lender indicates he's willing to make you a loan, there is nothing to worry about. You can also, in some cases, apply a spouse's income toward a loan.

The only problem is that lenders are authorized to lend certain percentages of the total package, from a low of 50 percent to a high of 95 percent or so. If you have ever taken a loan on a car, you'll remember that you had to provide a certain percentage of the total cost. Sometimes the trade-in of your old car would do it.

What percentage of the total package can you expect to finance? If you have a good potential for regular increases in pay in future years, and your past record is good, the percentage should be high. But with tight money, don't expect more than 80 percent. If you can make up the 20 percent with a cash down payment, you're "home free."

When applying for a loan, you'll probably have to specify the lot, type of house to be built, and the cost of other expenses, such as installing sewers or septic tank and surveying the lot. If you can get a separate loan for the lot, the job is simplified.

This all might indicate that we have the cart before the horse, and that you have to go get plans for the house, buy a lot, and have it surveyed and other things before applying for the loan. But if you can't get financing, there is no need to do anything else but wait for better times. Even if you don't have all things settled, you can see what your mortgage chances are.

Twenty-five to thirty years is the typical length of a mortgage. In addition to principal and interest, the payment is usually set up to include insurance and taxes. The insurance is usually life or disability to cover payments. One bright note: interest and taxes can be deducted from income for tax purposes. And it probably won't be too long before banks will, or will be required to, pay interest on any money they hold in escrow for paying taxes.

There are mortgage correspondents who can help you shop around for the best interest rate possible. Your lawyer can check. As for the lawyer's fee, you can agree on a percentage of the total cost, but you should have some idea of what it will be.

Don't expect to get a mortgage much lower than the local or area "going rate." Loans made at considerably lower rates are going to make up for it in other ways.

When you build, the mortgage starts out as a construction loan, which generally works like this: you pay interest only on the amount that you have borrowed, which grows as construction continues. This is also called a draw loan mortgage, and has a schedule of how many draws are to be made and how much each will be. Sometimes you won't know this schedule, so it could also provide for simply drawing money as each step of construction is made.

Another factor is the septic system and water. If the street is served by municipal water, no trouble. But if you have to dig a well, be sure to determine whether it can be dug on the same property that the septic tank (if no sewers) is installed.

A septic system is not just a matter of a hole in the ground. Just a hole in the ground is a cesspool, outlawed in most areas. A septic tank consists of a large concrete or steel tank (concrete is better), with an inlet pipe for sewage; solids collect in the tank and are broken down by microscopic organisms, and liquids flow out of an outlet into a leaching field. The solids are pumped out every two years or so, and the liquids seep harmlessly into the ground, well below the surface, which is unaffected. Leaching fields can be fairly large, depending on the permeability of the ground.

Now, regulations may require the well to be, say, 100 to 150 feet from the leaching field. Even if your lot is quite large, your neighbors' wells and leaching fields (to each side and to the rear of your lot) can influence the location of your well and leaching field. The regulation on distance between two installations applies to your installations as well as your neighbors'. So it is possible that your lot cannot accommodate both septic system and well.

Usually, a lot in a built-up area has water service. And if you have that and sewers, there is no worry. And out in the country, the lots, if indeed they are termed lots, are big enough not to cause a problem.

If you need a septic system, a percolation test on the property is a must, usually done by the municipality to determine whether sewage liquids can seep into the ground properly and fast enough. If your property is clay, percolation will be slow, and a septic system might be impossible, and disallowed. If the earth is sandy, or otherwise permeable, the leaching field's size is determined by how fast the liquids will percolate into the ground. If you have ever heard of a neighborhood of new houses having trouble with their septic systems, it may be that there was breaching of the percolation law, and a lot of people are stuck with overflowing leaching fields, the result of unscrupulous contractors and municipal officials.

Most septic systems are not excessively expensive. The cost of a tank is standard, and pipes in the system, plus excavation, don't amount to much.

Digging a well is another matter. There is no guarantee that a digger will hit water, or enough of it for the proper pressure, but I know of no one who has had a well drilled and come up with nothing. Water drilling, however, costs some $10 a foot, and if you go down 300 feet you'll quickly see where the money goes. If the lot is ideal in all other ways, then the expense of drilling a well and finding water makes it all worth it.

Then there is zoning, which was designed originally to protect the homeowner from having a machine shop for a neighbor, or to restrict an area, such as "Residence A," to one-family houses. Areas zoned "Industrial" or "Business" are quite obvious, so anyone buying a lot in such an area should know what to expect.

However, modern zoning laws sometimes seem to be against the little guy. For instance, "snob zoning" may keep you from an area that seems ideal. Snob zoning specifies lots of one or more acres each, perhaps, and you might not be able to afford that much land. Another form of snob zoning restricts the style of a house. If you want to build a modern style house in an area zoned for "early American," you simply cannot. Still another type of restriction is based on size. If you want to build a modest house of 1000 square feet of floor space, you may find that zoning requires a minimum space of 2000 square feet.

And zoning laws can change. If you buy a lot today with no capricious zoning regulations, and are not ready to build for two years or so, by that time restrictions may have been voted in. If this is the case, you can apply for a nonconforming use. This means that you want permission to build a 1000-square-foot house in a zone of 2000-square-foot minimums. You can get a sympathetic ear, and vote, if you tell the board involved that you bought the property two years ago with intent to build within three years, but cannot live up to the new restrictions.

When you go before a zoning board or other officials, be armed with your arguments, but don't be argumentative. Be also armed with drawings and other plans that are reasonably accurate and clear. If you can get the neighbors to support you, so much the better. Another positive point is whether you plan to occupy the house, as opposed to renting it out or building it for speculation.

Zoning laws also protect you. They keep neighbors from building shacks and living in them without benefit of permits and/or proper installations. And they keep you from building a palace in a neighborhood of shacks.

Sometimes you'll see a multifamily house or a store in a "Residence A" neighborhood. How did they get there? They could have been allowed through a variance, obtained by petitioning a zoning board. Politics often enters into such variances, too, and if these nonconforming uses seem widespread, the neighborhood might not be for you. More likely, the nonconforming uses probably were in the neighborhood before zoning went into effect banning them. These are "grandfather" nonconforming uses, which you may have heard about. Every neighborhood has some.

If you've ever owned a house or lived in an apartment you've probably bought homeowner's insurance, which covers the house and its contents against fire, wind, vandalism, theft, and many other losses. An apartment dweller's policy covers furnishings and personal items. There is considerably more insurance involved when building a house. To make sure you are covered properly and completely, use one insurance agent. The one who holds your present insurance will do. This way, one agent handles all policies and there is no dispute over which of the many policies should pay for loss.

Consult with this agent. He'll be glad to get your business and, if he's good, can save you money. Most laymen know very little about insurance except that it is supposed to pay in case of loss. If you tried to figure out the maze of policies involved in building, you'd have to go to insurance school.

Briefly, here is what is needed:

1. A builder's risk policy. This is much like the regular homeowner's policy that lasts the life of the construction project. One compensation is that the life of the policy is relatively short so it is not too expensive. If you specify a deductible clause you'll save.

A theft policy would be too expensive to have. Theft is always a big problem on a job, and it is difficult to catch such thieves. So the best protection is to make the project as fast and steady as possible, and to keep a minimum of material on the job. A huge pile of 2 x 4s, another pile of 2 x 8s and 2 x 10s, still another pile of luscious-looking spruce boards, plywood, and finish material like shingles and clear pine, is a tempting target.

So limit the materials on the job to those that can be used in a short period of time, like a few days. That means more deliveries of material, but with a big order the supplier should be glad to oblige. When the house is "to the weather," that is, can be locked up, with windows in, etc., a lot of the materials can be stored in the basement or in the house itself. A sign saying "Call Police" or "Dog on Premises" will discourage all but the determined professional rat. And a real dog on the premises might even get a rat or two.

2. Workmen's compensation. This covers workers who are hurt on the job. Without it, the builder-owner can get socked with a suit more costly than the entire project. Before work is started a builder-owner must get a workmen's compensation certificate from his insurance agent for each subcontractor (who will cover his own crews) or for each worker if he works alone. Sometimes a contractor will have the certificate. Just make sure workmen's compensation is in effect one way or another.

3. A liability policy. This will cover you while working on the house. It will cover any relatives, friends, or acquaintances who help you on the job. And it might cover a weekend crew of friends who get together with a couple of cases of beer to paint the house in one fell swoop.

chapter 3

It's All a Matter of Taste — and Budget

Plans and Styles

No matter what kind of a house you want: traditional, Colonial reproduction, or ultramodern, you'll find plenty of plans around at a minimal cost. And remember, the style will dictate cost and ease of building.

You may have seen plans in newspapers; every metropolitan newspaper has them, many smaller ones do too. The newspaper I work for carries a weekly house plan, and in some ten years I've never heard a contractor complain about them, although I've heard plenty of other complaints. This type of plan costs less than $20; quite inexpensive, which is a consideration since you'll need a set of plans for yourself, the contractors you hire, and possibly one for the community.

You can often buy a book of plans for a small investment, which gives you a chance to look them over before deciding what style you want. Or try the Yellow Pages. Look up an architect to see if he offers stock plans. They can be available for $100 and may be altered by the architect for little more. But if you have them altered, make sure they are redrawn to the scale of the original plans; otherwise they can give much trouble to contractors and cabinet installers.

If you like the idea of "traditional," you might consider a Colonial reproduction, or easier, a house in the Colonial style. A Colonial style house, in my opinion, has more style than a straight traditional. Both Colonial types and traditionals (ranches, split levels, etc.) are easier to build than modern, particularly ultramodern.

Colonials and traditionals have certain advantages. You can get a big, square Colonial, with a center entrance and center chimney; or a garrison, a rectangular job with two stories, with the second story jutting out a foot or so above the first story, usually in front. Then there is the so-called Dutch-Colonial, with its gambrel roof. That's the type of roof that has a double pitch: the lower part of the pitch is very steep, and starts at the ceiling of the first floor. The upper part of the pitch is quite shallow. "Eye" or shed dormers provide windows, along with those at the gable walls. This style is a variation of the classic Cape Cod, which has a regularly pitched roof starting at the first floor ceiling. Eye dormers permit windows in the long side of the second floor, in addition to those at the gable walls. In both Cape and Dutch-Colonials, shed dormers (with a slightly pitched flat roof like a shed, hence the name) provide more room along the long edge of the house.

The Cape and the ranch (traditional) are

probably the easiest for the layman to put together himself. In both, the roof starts at the first-floor ceiling, so you don't have to worry about going up to terrifying heights. A gambrel roof is pretty tough to build for a layman, because the rafters are set at two angles.

Many large two-story houses have a hipped roof; that is, with the roof sloping down to the eaves on all four sides. So to set the rafters of a hipped roof, you have the added difficulty of sawing them at two angles.

There is little saving in construction costs of a house due to style, except you can save in a traditional or Colonial house as opposed to an ultramodern one.

Because it is one story, a ranch house might give the impression of saving money, but any savings are eaten up by the greater foundation area. A big cellar is fine, but if you have a water problem, it is particularly frustrating to have a big cellar that can't be used because it's wet.

A modern house with long, sweeping roof lines, soaring ceiling, and huge exposed beams, is quite expensive to build. Not only does it require tricky construction techniques, but the oversized beams are more expensive than regular beams per board foot. Heating a high-ceilinged house costs more, too.

Most ultramodern houses are built with the post and beam technique, involving oversized timbers as well as double-thick roof decks, all adding to the cost of the house.

You may want to expose beams, or build a "cathedral ceiling" in the living room, dining room, or family room. This will add to the cost, particularly if they are true beams. A high ceiling, either ridged (rising on two slopes to one ridge) or shed (rising to a high plane on one side of the room), should be built on one story. That is, if you have a two-story house, plan for such a high-ceilinged room to be an "ell" or wing, with its own roof, so the ceiling can go all the way up to it. A family room wing might solve this problem.

If you like the idea of exposed beams in a flat-ceiling room, consider fake beams. There are plenty on the market, including real wood (hollowed out to form a U to save weight) and polystyrene. The polystyrene (Styrofoam) really looks like old beams but it would be better to opt for the real wood. If you like the idea of exposed beams in a room that has a room above it, and you want real beams, consider the fact that the floor of the room will be virtually exposed to the room below, and sound penetration will be a serious factor. That is why we think fake beams of real wood are best. As for sound control, that is covered in Chapter 18.

All things considered, the garrison Colonial house is probably the best buy when building yourself, with no fancy techniques that might throw the amateur builder. So that's the style of house (Figures 1, 2) we'll be talking about through this book.

It's got a lot of room, will look good with a "wing" built later, including garage and/or family room; the roof line is straight and simple, and the fireplace can be built in the center or on an outside wall. Of course, we will talk about such things as hipped roofs, valleys, dormers, and other building tech-

FIGURE 1. The rendering of a house (a garrison) is an idealized portrait of the planned building, designed to put the builder in the right frame of mind to get started.

niques, but with a garrison you won't run into them.

Another thing to consider in the style and design of the house is that you won't be living in it for the life of the house. Well constructed, it will last for one to two hundred years. In fact, if the wood is kept dry and protected, it could last a thousand. Even if your kids inherit it, they might not be living in it. So don't overdesign; that's costly. Be careful of built-ins you make just for you and your family; they may make the house difficult to sell.

The style of a house can be altered by using different kinds of coverings, windows, and doors. For instance, siding can be contemporary: wide clapboards or very deep shingles (12 inches or so) or vertical boards of almost any treatment. Or, siding can be traditional: narrow clapboards or shallow shingles (4½ to 6 inches).

Double-hung windows (those that rise vertically) can be contemporary, with horizontal muntins dividing each sash in half. Or windows can be Colonial, six to twelve panes in each sash, which can go nicely with any traditional design. (When you read of windows that are 6 over 6 or 12 over 12, it indicates the number of panes — lights — in each sash.)

Doors can be Colonial or traditional (deeply paneled) or modern (flush, with or without molding) or even vertical or diagonal boards held together with a Z brace. The panels in a traditional door can be square or rectangular or a combination of both.

Styles can be mixed, but it's not generally accepted to have a modern window and door treatment on a Colonial style house. There's no arguing taste, but a good example of how horrible mixing of styles can be is a perfectly good Colonial with a bunch of double-hung windows that have vertical panes in them. This often happened around the turn of the century when people thought "modern" 2 over 2 vertically paned windows were just the things to replace those "old-fashioned" 6 over 6 Colonial windows. The result was ugliness.

Another example of misguided "improvements" was the tearing out or bricking up of perfectly good fireplaces. Admittedly, fireplaces take up a lot of wall and floor space, but in a genuine Colonial or Victorian house,

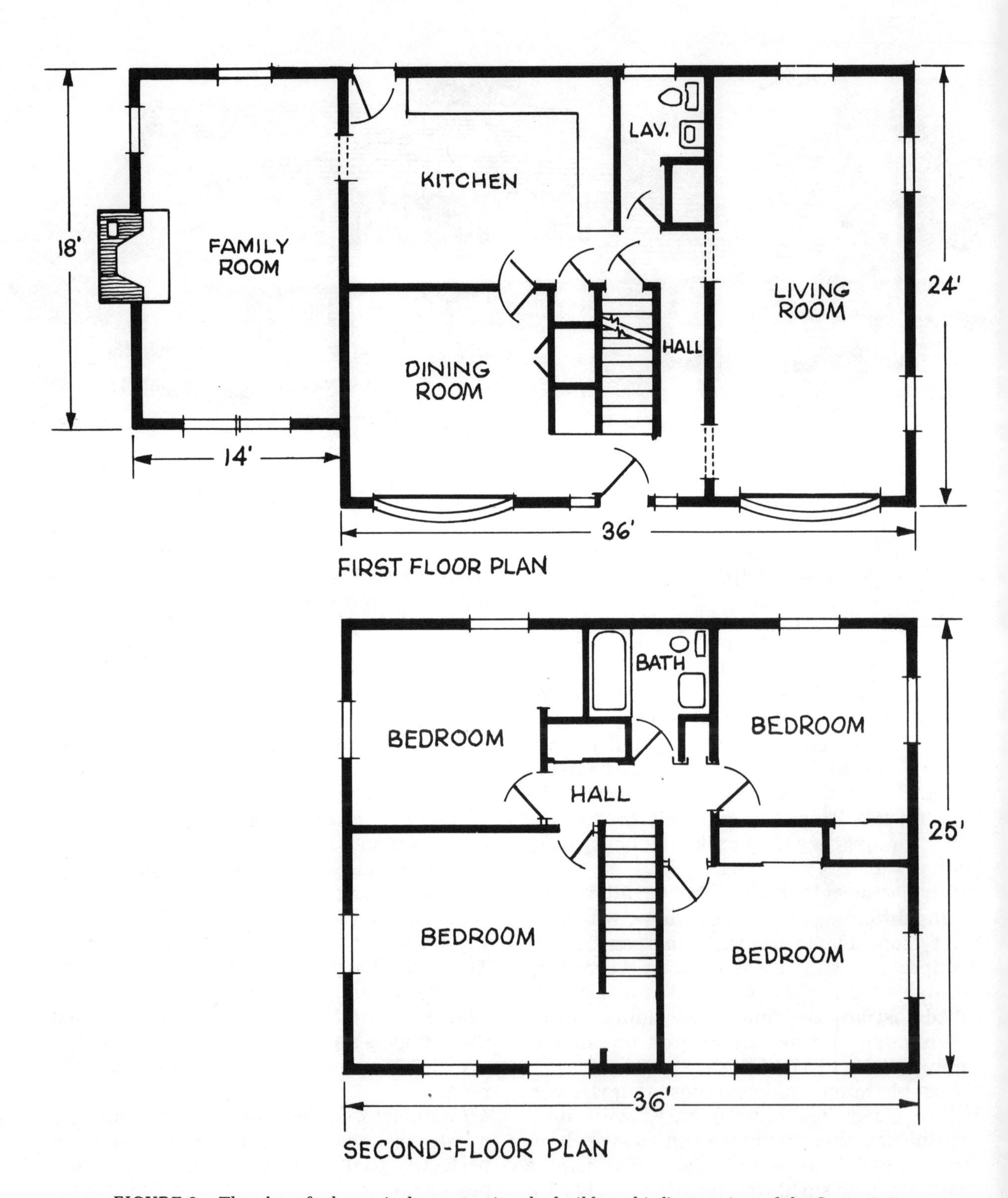

FIGURE 2. The plan of a house is drawn to give the builder a bird's-eye view of the floors; it is much more detailed in working drawings.

FIGURE 3. Ranch, all one floor.

it's an architectural crime to remove basic parts of a house's design, particularly if they cannot be restored without being replaced.

Other treatments can make a house traditional or modern:

Traditional: narrow eaves, wooden porches with balustrades and pillars, chimney treatment to imitate Colonial or European styles, a cupola on a low roof of a ranch, wing, or garage, and trim boards that come plain or molded.

A good traditional chimney style is the Tory, painted white on the bottom with a black band on top. Legend has it that Tories painted their chimneys in this manner to identify themselves as loyalists who would receive preferential treatment from the king's soldiers.

Contemporary: very wide eaves, low-sloping and very broad roofs, plain but often wide trim, wrought-iron railings and pillars; massive chimneys, wood decks and unusually shaped wings, porches, and decks.

Let's look at some of the house styles and their advantages and disadvantages.

1. Ranch (Figure 3). Everything on one floor, on a slab or regular cellar.

Advantages: ease of construction; no great heights to worry about; heating and cooling systems are simple to install, economical to operate; no stairs to climb; easy to expand, either with a wing or second story.

Disadvantages: reduced privacy; may be difficult to locate on a small lot; foundation cost high due to size.

2. Raised ranch or split entry (Figure 4). A ranch on a cellar that is about 50 percent above ground level; often uses the cellar as a regular living floor.

Advantages: same as the ranch, plus a basement that can be used readily for expansion.

Disadvantages: same as a ranch, plus difficulties of basement water seepage.

FIGURE 4. Raised ranch, or split-entry. The basement is shallow, with half a stairway to the basement and half to the top floor; it can be used as a living or expansion room.

3. Cape Cod (Figure 5). A classically easy design to build, where the second floor starts at the roof line, with windows at each gable wall and optional "eye" or "A" dormers foı light and ventilation. Capes come in half, three-quarter, and full sizes.

Advantages: easy to build. Roof line is low. Rooms can be finished on first floor and left rough on second floor for finishing later. Second floor, with a shed dormer for more room, can be two large bedrooms, a youngster's hideaway. Shed dormer can be built later.

Disadvantages: space on second floor is restricted (but solvable by dormers) by the

FIGURE 5. Cape Cod. Roof line starts at the top of the first floor. Eye dormers in front are for light, ventilation, and decoration; shed dormer in back is for added second-floor space.

slant of the roof. Eye dormers are tricky to build.

4. Dutch-Colonial or gambrel roof Cape (Figure 6). The gambrel roof has a steep pitch on lower part, a shallower pitch to the peak. Many barns have a gambrel roof.

Advantages: all those of a Cape, with the gambrel roof allowing more room than a Cape.

Disadvantages: the roof is tricky to build, but because of the steep pitch the rafters do not have to be as heavy as on the Cape or other straight-pitched roofs.

5. Garrison (Figure 1). Two full stories with a regularly pitched roof, and a 1-foot extension of the second floor above the first, usually in the front. The blockhouse fort of early American days has this type of extension, usually on all four sides.

Advantages: very roomy, with slightly more floor space on second floor. Easy to build, with no specialty construction such as dormers.

Disadvantages: roof line is high, causing problems for the amateur.

6. Colonial (Figure 7). A general term for a house with two full stories, a gable roof line or a hip roof (one with the roof pitch coming down on all four sides).

Advantages: very roomy.

Disadvantages: high roof line. Hip roof is difficult to build.

7. Split-level (Figure 8). Two or more floors, with the upper floor usually only half again as high as the first. Kitchen, dining room, and living room usually on one level, the ground level; bedrooms on another, half a story higher. The ground level is usually over a full basement; the half story over a garage on grade.

Advantages: adapts well to a lot with a slope. The split can be front to back, back to front, or to one side, depending on the slope of the lot. A split-level on a flat lot is not an ideal design.

FIGURE 6. Ranch with two-story gambrel-roofed wing. A gambrel roof has slopes of different degrees on each side of the ridge.

Disadvantages: more complicated to build. If the garage is built into the house, fire walls must be built of masonry, and there's a problem of auto exhaust. Drivers must get into the habit of moving the car out of the garage immediately and not closing the door too soon. And when the car enters the garage, the engine should be stopped immediately. This practice is wise anyway,

FIGURE 7. Colonial, generally two stories, with center entrance and chimney (the chimney also can be at one or both sides), and a hip roof (with roof slope going down to the eaves on all four sides).

whether the garage is attached or not. There are several roof lines which can cause problems for the amateur.

8. Modern (Figure 9). Anything goes, from long sweeping roof lines to half a dozen levels. Sometimes post and beam construction is involved.

Advantages: imagination takes over here; there are no rules to follow. Gives good opportunity for cathedral ceilings, exposed beams and large rooms, balconies, and other innovations.

Disadvantages: plans can be expensive and so can building, with many individual cuts required in wood beams. The larger the beam, the more expensive it is per board foot. For instance, a 2 x 12 may be more expensive per board foot than a 2 x 6. And an 8 x 12 or 8 x 16 beam is proportionately more expensive than a nominal 2-inch beam.

9. Saltbox. A two-story Colonial type with a short pitched roof in the front and a very long pitched roof at the back, descending to the first-floor ceiling or even lower. Named for the old-fashioned wooden saltboxes with the same contours.

The idea of this design is twofold, according to history. One legend has it that the long, broad roof line was built to face south to take advantage of the warming sun in the winter. The snow load also acted as insulation. The other legend is that when wily Colonists and early Americans added to their house they made the long roof line in the back so the assessors wouldn't see it and add to the taxes. Well, it was a pretty lousy assessor who didn't notice the change in the roof line.

Advantages: attractive, easy to build, and with a fair amount of extra space in the long, sweeping area in the back.

Disadvantages: plans are not common.

FIGURE 8. Split level, with at least three levels of living, or utility space, without a full stairway.

FIGURE 9. Modern, the sky's the limit in design.

The long roof line may pose problems for the amateur.

Suppose you want to add to your house sometime in the future.

The traditional style is simple to add to, from a "good looks" point of view. A ranch can be added to by putting a second story on top, or by building an "ell" on the ground. A garrison or Colonial can be added to with little difficulty, either on the side or at the back, depending on lot space. A Cape might pose problems, because of the difficulty of tying the roof into the roof or siding of the original building. This also applies to ranches. Adding a wing to a gambrel roof poses the same problem but don't worry about putting a straight-pitched roof on the wing of your gambrel roof. It's done all the time and is an accepted practice. And even if it isn't "accepted," you can do what you want to do. Adding wings to split level and modern houses is restricted only by your imagination.

chapter 4

The Stuff a House Is Made of

Materials: Their Characteristics and Uses

In the back of this book is a glossary of terms, some of which may be familiar to the amateur, others completely unknown, and still others "way out" or at least very strange.

But it would be helpful here to describe some of the materials used in a house.

Starting at the bottom, let's consider concrete in the foundation. It is a plastic material made of Portland cement, a baked and ground limestone named for Portland, England, and mixed in various proportions of cement, sand, aggregate (gravel or crushed stone), and water. Aggregate also can be materials like mica, for lightness in weight.

Concrete cures (hardens) into a very solid material that is very strong in compression (stress from top to bottom) but not strong in tensile strength (the ability to resist forces trying to tear it apart). To increase its strength, concrete is poured around reinforcing rods (steel rods with little bumps on them, to promote the gripping power of the concrete). One of the funny terms in building is the term for this steel: deformed rods. Other reinforcing materials are steel mesh, particularly usable for reinforcing floors and slabs.

Concrete blocks are often mistermed cement blocks (just as concrete is mistermed cement) or cinder blocks. They are made of cement, sand, and water, and are not quite as strong as poured concrete. Cinder blocks are made from cinders as an aggregate, designed to make them lighter. They are more expensive than concrete blocks and are less water-resistant.

Concrete, by the way, is not waterproof by itself, although it is water-resistant. It is made waterproof by certain additives, increasing its cost, and also by backing it with a vapor barrier like polyethylene plastic sheets or coating it with tar.

Lumber is the basic frame of a house and includes 2 x 4s, 2 x 6s, 2 x 8s, and 2 x 10s, and some planks or beams such as 4 x 6s, 6 x 8s, etc. They are not full dimension, although they are called dimension lumber. A 2 x 4, for instance, is not 2 inches thick and 4 inches wide, but rather 1½ inches thick and 3½ inches wide. The reduction is due to dressing (planing) the lumber from a rough (full-sized) state to its dressed state. Generally in dimension lumber you can take half an inch off each dimension. The greater the width, the farther away it is from the full dimension. A 2 x 12, for instance, is 1½ inches thick but only 11¼ inches wide.

Other building lumber includes boards, ranging from 1 x 2s to 1 x 12s. The 1-inch thickness becomes ¾ inch dressed. The other dimensions are similar to those of dimension lumber.

Tongued and grooved boards, used for subfloors and sheathing (the outer wall covering beneath the siding) are reduced even more in width, because the tongue on one edge is included in the dimension, but when installed the tongue is covered. So a 1 x 6 tongued and grooved board is closer to ¾ x 5 inches than ¾ x 5½. Tongued and grooved boards are also used for paneling and finished flooring, and are called matched boards. When the tongues and grooves are along the sides and the ends as well, the boards are called end-matched.

Lumber is sold by the board foot, which is a piece of wood nominally 1 inch thick and nominally 12 inches long and 12 inches wide. A 1 x 12 12 inches long is 1 board foot. A 2 x 12 12 inches long is 2 board feet, because of its double thickness. Lumber dealers use formulas to break down the amount of 2 x 4s and other dimension lumber into board feet.

Highly popular and a real asset in construction is plywood. It comes in many thicknesses, but its dimensions are exact. Most plywood comes in 4 x 8–foot sheets, although some can be ordered in 10-foot lengths and longer. It is sold by the sheet or square foot.

Plywood is made of several plies of wood, hence its name, laminated together with glue. The term exterior plywood means that its glues are water-resistant. There is no truly waterproof plywood, because wood is not waterproof. The nearest thing to a waterproof plywood is marine plywood. All plywood should be painted if it is exposed to the weather. In construction, it's best to use an exterior-grade plywood except for fancy trim and other surfaces.

Another wood material with exact dimensions is millwork, or moldings. It is sold by the running foot, which is 1 foot long no matter what the other dimensions.

Siding, such as shingles and clapboards, is made of white or red cedar and sold by the square, which is 100 square feet.

Roofing shingles are made of felt and asphalt bonded with crushed stone to make them hold together and resist weather better. These are also sold by the square and, as in the case of wood siding, are broken into bundles. Asphalt shingles are graded by weight. A 235-pound shingle weighs 235 pounds per square, installed. Roofing felt, or "tarpaper," is made of asphalt-impregnated paper or felt and weighs 15 to 45 pounds per square. Roll roofing is made just like asphalt shingles but comes in rolls, and is used on very shallow-pitched roofs.

Plasterboard is a sheet of plaster bonded on two sides by paper, and comes in various thicknesses and sheet sizes, from 4 x 8 feet and longer. Water-resistant plasterboard is used in areas subject to dampness or where water is used.

Other materials are described wherever they are listed in construction.

The materials need putting together in a

house, and the nail is the most popular fastener. Here is a rundown on some of the kinds of nails used in construction:

Common: wire nails, bright (ungalvanized) and galvanized steel, with a large head, used in rough framing. Galvanized nails of all types are more expensive than bright steel but are better because they resist rust and hold far better. One thing we've found is that when you use galvanized, make sure they are the hot-dipped zinc type and not the electroplated type. Because galvanizing adds to their thickness, use box nails. Although thinner than common nails, when box nails are galvanized they become approximately the same thickness. We say this because too thick a nail can cause splitting of the wood. Also, galvanized nails hold the wood much better than brights.

Box: like common except thinner, designed to be nailed into the thickness of nominal 1-inch boards.

Ring-shanked: nails with their shanks grooved in a series of unconnected rings. They hold very well, and are used for nailing underlayment plywood, hardboard, and particleboard as a base for tile, linoleum, and carpeting, and for nailing materials like insulation board, which are not as rigid as plywood or lumber. These nails also help hold plywood to floors and walls; in fact, sometimes they're so strong that they cannot be pulled out without destroying them or the wood they're holding.

Screw: nails that are spiral-grooved much like screws, and they actually turn when they are driven. Their heads are not as big as those on common nails, so they can be countersunk and covered with putty when applied to a finished floor.

Plasterboard: ring-shanked, large-headed nails for securing plasterboard to wood. Many nails are named for their use or the materials they secure.

Cut: square-edged; made of hardened steel bars. They have square-edged points as well, so they can be driven into materials like oak floorboards. The dull point actually crushes the wood fibers instead of separating the fibers as a sharp point will, and reduces splitting.

Roofing: large-headed galvanized nail, designed to secure asphalt shingles to wood. Can also be used to secure roofing felt and other sheet goods when stapling is not recommended or practical. They also are recommended for nailing the vinyl flange of vinyl-clad setup windows.

Shingle: thin, galvanized nails for securing wood shingles. Clapboard nails are not called that; they are simply a medium-sized galvanized box nail.

Finish: headless or nearly headless nails for securing finish wood interior materials like molding and window and door casing. These can be countersunk and filled with wood putty. It is not generally recommended that finish exterior materials be secured with finish nails because of the size and weight of exterior trim. Instead, casing nails, with a medium-sized head, are recommended, and can be countersunk and filled with putty.

There are a million, or what seems like a million, other kinds of nails on the market, but the ones we have listed are about the extent of those you'll be using when building a house.

One specialized nail you will be using, however, is the scaffolding, or double-headed, nail. It is used for temporary nailing: scaffolding, braces, etc. The double head allows the nail to be driven up to the first head, making a secure hold. The second head, fully exposed, allows easy removal.

Now, about size. Common and finish nails come in sizes graded by penny (abbreviated d), indicating length and thickness. The penny system was devised in Olde England to indicate the price per 100 nails. A 2-penny nail sold for 2 cents per 100. A 16-penny nail, much larger, sold for 16 cents per 100. The small "d" symbol for penny is derived from denarius, an ancient Roman coin, originally equivalent to 10 bronze asses. It's Latin, meaning "consisting of ten," which may have something to do with nails selling by the 100. At any rate, the penny sizes are listed here, and through the book the proper-sized nails will be included with each nailing job.

A simple rule to be remembered, but not followed doggedly, is that a nail should be long enough so that its length equals the thickness of the material being held plus 1½ to 2 times that thickness in the holding material.

Thus, when nailing nominal 1-inch board, the nail should go through the board (¾ inch), and twice that (1½ inches) into the holding material, for a total of 2¼ inches. This is the length of a 7-penny nail, but it is recommended that most nominal 1-inch lumber be secured with 8-penny nails, which are 2¾ inches long. So rules are made not to follow to the letter.

And when you go to nail nominal 2-inch lumber, the rule would require that you use a nail 4½ inches long (1½ inches for the material being held plus 3 inches into the holding material). A 30-penny nail is 4½ inches long, far bigger than the 16-penny nail recommended, which is only 3½ inches. But the 16-penny nails hold fully adequately. Anyway, here are the penny sizes:

4 penny, 1½ inches
5 penny, 1¾ inches
6 penny, 2 inches
7 penny, 2¼ inches
8 penny, 2½ inches
9 penny, 2¾ inches
10 penny, 3 inches
12 penny, 3¼ inches
16 penny, 3½ inches
20 penny, 4 inches
30 penny, 4½ inches
40 penny, 5 inches
50 penny, 5½ inches
60 penny, 6 inches

If in doubt when buying nails, specify the job to be done, and you'll get the right nail.

chapter 5

Asearching We Will Go

How to Get Permits, Materials, and Workers

You're not about to turn over that first spadeful of dirt, eager as you are to get started, without first getting down to the nitty gritty of getting bids for materials and work.

Now that you have the land and the house plans, plus a plot plan (how the house is situated on the lot), you can apply for a building permit.

A permit should not take long to obtain, and it should not be expensive, usually a percentage of the estimated cost of the project. Most communities are pretty good concerning permits; others can be pretty persnickety and can demand as many as thirty-two inspections on various permits.

Let's hope you won't run into that kind of jungle, but you will need such inspections as: footings and foundation before backfilling (filling in the trench on the outside of the foundation); when the house is "to the weather," that is, complete with roofing, siding, and windows, but no insulation and interior walls; finally, an inspection to certify occupancy.

Specialty inspections would include rough and finish plumbing, rough and finish electric, septic tank, and water well.

Some communities don't seem to give a damn and all you need is septic tank inspection and plumbing and electrical inspections; not always all of them, either. If you are in such a town, it's best to buy a code book of a nearby community or of the state because uniform state codes are coming into being, to which communities must adhere.

At any rate, if your community has a code book, buy it. It will be a good investment. Also, more and more communities are requiring permits for electrical, plumbing, and septic tank work to be taken out and the work done by licensed practitioners; therefore it would be difficult or impossible to do the work yourself, even if you were skilled at such work.

If you dig that sort of work, you might be able to find a subcontractor whom you could aid. But still, the finish work would probably have to be done by him and he may, in order to get inspection approval, have to certify that he did the work.

The rest of the work you should be able to do. All you need is the inclination and time.

Which brings you to the lumber list. The house plans have, or should have, a lumber list: how many 4 x 6 sills to go on top of the foundation, 2 x 8 or 2 x 10 floor joists (horizontal beams), 2 x 8 rafters (roof beams), 2 x

4 studs (upright wall beams), and plates (horizontal wall members); also, ⅝-inch exterior plywood for subfloors, finish flooring lumber or ½- to ⅝-inch particleboard for underlayment, ½-inch exterior plywood for wall and roof sheathing, gutters, clapboards, or wood shingles, asphalt or other type of roof shingles.

Woodwork for interior and exterior: 1 x 2s, 1 x 4s, 1 x 5s, 1 x 6s, 1 x 12s; baseboards, quarter rounds and other interior trim. Windows, usually "setup" windows that come in one unit and are simply inserted in rough openings. The same goes for doors.

Half-inch plasterboard for dry-wall application, or plasterboard lath for plastering, plaster, insulation.

Electrical material, plumbing material, ceramic tile, resilient tile and linoleum, bricks and other masonry supplies, steel or wooden beams. Even nails, screws, and other hardware, the type and number, are included, or should be.

Not all this stuff has to be bought at once. It will depend on how much you are going to do yourself. Which brings you to contractors. There are many kinds of contractors, which we will list below. How many of these contractors you hire will depend on how much of the work you will do yourself. We can't decide that for you, so we will list not only the jobs to be done in future chapters, but also the type of contractor available.

But let's go back to the lumber list. If your house plans do not include one, here's how to obtain one. Go to a reliable building supplier and show him the plans. He should be able to "take the specs [specifications] off"; that is, determine from the plans pretty well what you'll need to build the house. Then he will give you a bid. Have an exact copy made of the first lumber list, and submit it to other material suppliers for their bids.

When buying materials or labor, it's best to get at least three bids. You can compare each bid and determine which is best for you, and most economical. The lowest bid may not be the most economical; a low bid may be due to skimping on the amount or using lower-quality materials. For instance, a bidder might specify 2 x 3s for certain walls instead of 2 x 4s and not tell you.

For framing lumber, Douglas fir is best, but a combination of hemlock and fir is entirely adequate. For woodwork in and out, pine is best. If it is clear pine (without knots and other blemishes and defects), it will be much more expensive than No. 1 common. Now, No. 1 common is every bit as strong as clear pine, but here's the rub. If you are going to paint the woodwork, in or out or both, you have to seal the knots with shellac, aluminum paint, or a stain killer, or the knots will show ugly marks through the paint. It will be very frustrating, and there is no guarantee that the shellac, aluminum paint, or stain killer will keep those marks from showing through. If you are going to stain the woodwork, in or out or both, then the knots will make little or no difference, except they will add a "rustic" look to the trim. Clear pine is definitely the best stuff to get, but you may be shocked at the difference in cost: 50 to 100 percent higher than No. 1 common.

So, know what you're getting from a materials supplier. And when you get three bids for the same amount of supplies, say, for $14,000, $10,000, and $9,950, and can determine that the materials offered are similar in amount, size and quality, you can be safe in eliminating the high bidder. By the same token, if you get three bids for $10,000, $9,950 and $6,000, everything else being equal, you can be safe in eliminating the low bidder. Such great discrepancies mean that the high bidder doesn't want to sell you the stuff but wants to make a bid anyway, or he's an unscrupulous dealer; and that the low bidder wants the bid desperately and may be in financial straits.

So how do you contact and obtain bidders and contractors? The Yellow Pages are a good source. So are newspaper advertisements, usually in the classified section of the Sunday editions. There are various categories, but usually something like "Services and Repairs" or "Market Basket" is where they are advertising.

Cruising the neighborhoods can be a help, too, looking for houses under construction. Talk to neighbors or friends whom you know have done such work. Word of mouth has a habit of spreading your word very nicely. Call the home and garden editor of a nearby metropolitan newspaper; he or she is usually fairly knowledgeable and sometimes might be willing to give you a name or two.

How can you determine if such people are reliable? Well, most building suppliers in an area have been there for a while. In fact, local lumberyards and building supply stores may be family-owned outfits, having been in business for generations. If a contractor has been at one address, or at least in the same community, or nearby, for five years or more, you can be fairly confident that he hasn't been run out of town on a rail.

Call the Better Business Bureau or the local Chamber of Commerce. Usually such bureaus will have information on a contractor only if he has had complaints filed against him. Then again you might get lucky and find compliments filed with him, too. Again, talk to people who have had work done by a contractor. Of course, the contractor himself might give you only satisfied customers. If you can obtain a list of people he's done work for from another source, so much the better.

Common sense and gut reactions have a lot to do with a successful contract. Get contractors as close to home as possible; that will cut down on expenses and you can find customers — satisfied or not — of those contractors more easily.

When you hire a general contractor, he does all the work: ordering of materials, hiring of workers and subcontractors, and the worrying. He'll probably do the framing and sheathing, siding, and roofing shingles. Insurance and other incidentals are also his concern. All you do is pay him. And general contractors usually work at a 15 percent profit margin. So if you do your own general contracting, you might save yourself that much as a starter.

If you are your own general contractor, here are the subcontracts you'll be dealing with:

1. Excavation

2. Concrete footings and foundation
3. Steel or wooden beams and columns for holding the entire house up
4. Framing. Installation of all dimension lumber: floor joists, wall studs, roof rafters, ceiling joists. Also, sheathing of walls, floors, and roof, installation of siding and roofing shingles, installation of windows, exterior doors, and exterior trim
5. Masonry for chimney, fireplaces, porches, and steps
6. Plumbing, including fixtures
7. Electricity
8. Insulation
9. Dry wall or plastering
10. Kitchen cabinets
11. Interior finish (trim), including baseboards, window, and door casings
12. Painting and wallpapering
13. Ceramic tiles
14. Linoleum and/or resilient tile
15. Exterior stain and painting
16. Grading
17. Landscaping

By looking at this list, you can determine how much you want to or can do by yourself.

At any rate, a casual look at a large city's Yellow Pages reveals these people and outfits offering their services and materials:

Architects
Bricklayers
Building Contractors
Building Materials
Buildings, Precut, Prefabricated, and Modular
Burglar Alarm Systems
Carpenters
Cesspools
Chambers of Commerce
Chimney Builders
Concrete Blocks
Concrete Construction Forms and Accessories
Concrete Contractors
Concrete — Ready-Mixed
Contractors, General
Copying and Duplicating Services
Dry Wall Contractors
Electrical Contractors
Electrical Supplies
Excavating Contractors
Floor Laying
Floor Materials
Furnaces — Heating (generally hot-air systems)
Heating Apparatus (generally hot-water systems)
Heating Contractors
Insulation Contractors
Iron Work (steel beams)
Kitchen Cabinets and Equipment
Landscape Architects
Linoleum
Lumber
Mason Contractors
Nurserymen
Painting Contractors
Paper Hangers
Plumbing Contractors
Rental Services
Roofing Contractors
Septic Tanks
Siding Contractors
Tile — Ceramic Contractors
Tools — Rental

Vacuum Cleaners
Well Contractors
Windows
Wrecking Contractors

A subcontractor will probably provide his own materials, and these are included in his bid. In some cases, the sub will work with your materials, but this is generally true only of framing: you buy the lumber and other supplies and he will put it together.

In the Yellow Pages list, I have included some services that might not seem appropriate, but wait. Duplicating and copying services can copy your plans, and more important, your lumber list, for bidding purposes. Rental services are important if you do a lot of the work yourself. Tools are treated in Chapter 6.

Wrecking companies are important to consider, too. For instance, you might be able to get from a wrecking company dimension lumber (nominal 2 inches thick and thicker) plus heavy beams for 30 to 40 percent less than new lumber. The lumber is good. The only thing to beware of is sawing through hidden nails. A contractor might object to that unless you paid for resharpening his saw blade every time he hits a nail, because once a saw hits a nail, it's useless until sharpened. If you do the work yourself, it's a risk worth taking. Just lay in a supply of saw blades so that when you dull one, you'll have several in reserve until the first one comes back from the sharpener.

All things considered, used lumber is a good risk for the price, and it is usually well seasoned, having been long dried out, if it has been kept under shelter. If so, it won't shrink in service; that is, will not shrink after it has been installed.

One thing about ordering the lumber you plan to use, primarily from the lumber list or taken off the specs — it's a good idea to make the deal with the lumber supplier as soon as possible, because in these days of superfast price hikes, the sooner you agree on a price for lumber and sign a purchase agreement, the lower the price is likely to be. Then you can order the lumber delivered as construction progresses. It's like today's lumber at yesterday's prices.

chapter 6

Bare Hands Just Won't Do

Tools of the Trade

Let's suppose you want to do most of the work yourself in building a dream house. That is what the rest of this book is all about. But those who may not do much of the work can find the remainder of the book useful because by reading it, and rereading it, they can get a good idea of what goes into house building and can determine whether the general contractor and/or subcontractors are doing a decent job. Or at least ask intelligent questions.

In other words, you'll be able to tell the difference between a joist and a rafter, and the subs will know it, and perhaps not try to put one over on you. This is not to say that most contractors are dishonest, but your knowledge will keep them on their good behavior.

So if you're going to do it yourself, a few words on equipment, tools, and helpers.

Basic tools go without saying, so we won't say it. But hammers and other basics should come in pairs. One for you and one for your spouse, maybe, or one for your helper, if he is so square as to show up on the job without one. Same goes for saws, but for a different reason. When one gets dull, send it to get sharpened and use the spare while you're waiting.

Here is a list of not-so-common tools that you'll find very handy.

1. A four-foot mason's or carpenter's spirit level, made of wood or metal. It will do wonders in making your work level (horizontal) and plumb (vertical). The longer the length, the more accurate the level.
2. A line level, which is a little spirit bubble with two hooks on it that you hang on a string to make sure extra-long spans are level. This is important in laying out the foundation.
3. Surveyor's string, stout twine that not only does not readily stretch, but will not sag when pulled tight.
4. A 50-foot tape measure for measuring foundations and other long lengths and spans.
5. A 12-foot tape rule. If you buy it an inch wide you can measure up to 7 feet without support. Such a tape will not "break" or bend at right angles like a wet piece of spaghetti. A ¾-inch rule is okay, too, but a ½-inch rule is not much good at all.
6. Carpenter's pencils. Very thick, rectangular in shape rather than round, so you have to sharpen them with a knife.
7. Utility knife. A large handle holding a

razor-sharp replaceable blade that will cut anything from wooden shingles to roofing shingles and plasterboard.

8. Carpenter's square. A large steel square (the long arm, called body, is 24 inches long and the short arm, called tongue, is 16 inches long), also called a rafter and/or framing square. It is calibrated, and if you're a mathematical genius you can measure many angles for cutting, particularly rafters, which must be cut at an angle so they'll set properly at the ridge.
9. Combination square. A small square for making right-angle and 45-degree marks on regular lumber.
10. Adjustable square. This is a little tool, like a combination or L square but with an adjustable blade, which can be loosened or tightened by a wing nut. It's handy in determining angles other than 90 and 45 degrees.
11. Sawhorses. You can build a pair from 2 x 4s, or you can buy brackets and use them in making a pair of 2 x 4 sawhorses. A third type of sawhorse is one with a wide top for ease of setting boards and with a shelf for storing tools.
12. Masonry tools. If you're really ambitious and want to build your own chimney and fireplace, you'll need such things as a trowel and pointing tool, the latter being a modified S-shaped piece of steel used to smooth off or strike joints in brick.
13. Wheelbarrow. For the dozens of carrying projects on the job. For mixing mortar in.
14. Power tools. Most important to fast work in framing a house is a carpenter's or framing power saw, a portable circular saw. Get a heavy-duty one with three or four blades, because that saw will be busy for days. And when a blade gets dull (or hits a hidden nail for instant dullness), you can send it to the sharpener and use a spare. A table saw is another good tool, often used by professional crews but a major investment even when building just one house. You can do without one. An electric drill might be helpful for predrilling nail holes in areas where driving the nail directly could cause the wood to split. Also, when toenailing (driving the nail at an angle into the material to be secured), a predrilled hole can be better "aimed." A saber saw is good for cutting plywood paneling, but not so hot for ripping (cutting a board with the grain) or for ordinary cross-cutting (across the grain). You will find you can make more accurate cuts (where accuracy is the prime consideration) with a hand saw.

This list of tools can mean quite an investment. So if you can find a helper, father-in-law, or old buddy who has some of these tools, take advantage of his good nature and borrow them. Naturally, if you hire a helper, he will bring his own tools.

Rental tools and equipment are another consideration.

You'll probably want your own ladders, but on occasion you'll have to rent a telescoping ladder when more than one man is working in high places. Also, two ladders are essential to set up a ladder scaffold utilizing two 2 x 12s as a platform.

Scaffolding can be rented, and might be a good item to include in your budget. It is not expensive, because there is nothing to break or break down. No moving parts, that is. Scaffolding comes in pipe form, which you can put together into a regular maze of supports and platforms. It is essential to make the platform as wide and stable as possible, with even a railing for extra safety. Remember, when you're on a scaffold, never step back to admire your handiwork.

Now, about those helpers. In building a house, handling 16-foot rafters and joists, erecting stud walls, lifting beams, etc., an extra strong back and set of willing hands are essential. You've heard of "two-man jobs"? Well, here is where two men are worth more than twice the value of one.

For instance, if many measurements must be taken in a high place, pieces cut and placed, think of the number of times you'd have to climb the ladder, measure, climb down, go to the sawhorses and saw and return to the ladder and climb up it to install the piece, if you worked alone. With two men, one can measure and install while the other saws to size. A sawyer can keep up with the measuring and installing. The same goes for cutting rafters, studs, plates, joists, and plywood.

The two-man theory also applies to setting up a stud wall, putting rafters in place, mixing mortar and laying bricks, putting joists in place over a 12-foot span. When you lug a few heavy boards around and try to heave a half-inch piece of plywood 4 by 8 feet into place all by yourself, you'll get the point.

chapter 7

Getting Down to Earth

Excavation, Footings, and Foundation

Finally. It's time to get to work, and you're ready to excavate.

But here again you have to apply all the patience you can muster. First, test borings must be made to determine whether there is a rocky ledge under the property. If so, removal by blasting might be necessary.

You might encounter a high water table, requiring design change or extra amounts of waterproofing and drainage. In fact, no matter how dry the excavation seems, it's always wise to use a maximum of waterproofing and drainage.

The design change might require making a crawl space or even building the house on a concrete slab, with no basement at all, or half a basement. This might sound discouraging, but it would be better to build a crawl space or slab, which is cheaper, than to have a constantly wet basement, which costs more and would be useless.

A word about basements. Do you need one at all? Here are some reasons why you don't:

They are expensive.

If there is a water problem, you could be plagued with seepage for the life of the house.

They tend to be damp, if only by condensation of water vapor on the cold concrete, and are therefore impossible for storage of clothes, books, paper products, and metal. Dehumidifiers will be needed to keep a basement dry, particularly in the summer. These do work well, but are another electrical appliance to use up valuable — and expensive — electricity.

They are not convenient, making an extra set of steps to go up and down.

A house can be designed to provide garage and sizable storage space and/or utility room, plus a family room.

Now, here are reasons why you might be able to do very well with a basement:

Furnace and fuel storage is out of the way and not taking up valuable floor space in the living areas.

A recreation room is fine, if it isn't wet.

A split-level house needs only a partial basement, with the other part devoted to a ground-level garage or crawl space.

If your lot slopes, the basement would be a "walkout" basement, on ground level, at the low part of the slope. This would solve some of the problems of ventilation, dampness, and seepage.

In raised ranches, the basement is only about half as deep into the ground as full basements, reducing the likelihood of water problems and providing as much space as the floor above.

Generally, there are fewer basements and more crawl spaces and slabs in warm climates, and more basements in colder climates. And a basement does provide a certain amount of barrier against the cold, particularly if the basement ceiling is insulated.

To help you decide about a basement, check the foundations of neighboring houses. Neighbors are willing to talk about their problems, particularly how they solve them and how you can avoid them.

Once the lot is cleared, the location of the house on the lot must be plotted. A surveyor, beforehand, will have marked the corners of the lot. After the clearing, the rough corners of the house are established. This location must be approved by the community; in fact, the plot plan is one of the musts required for a building permit. The plot plan must adhere to community requirements, such as front yard setbacks and side yard setbacks, which are in the town code, and which also can be influenced by neighboring houses.

Now, after the corners of the house have been established, you must get the corners square and the grade level, and make a guide for the excavator.

To do this, drive small but sturdy stakes into the ground at the outside corners, with a nail driven into the top showing the exact corners, the outside of the proposed foundation wall. How do you make a square corner? There are several ways. A 4 x 8–foot sheet of plywood is square. You can also use this formula: a right triangle with dimensions of 3, 4, and 5. That is, measure one leg of the corner 3 feet or multiples of 3; measure the other leg 4 feet or multiples of 4; measure the diagonal connecting the two legs 5 feet or multiples of 5. The angle formed by the 3- and 4-foot lengths is 90 degrees. Another way is when the rectangle of the house has been set up, measure the diagonals of the rectangle; that is, from one corner to the diagonally opposite corner. When the two diagonals are equal, the corners are square.

If your house has a wing or ell or other extension, measure it and square the corners separately. Of course, things like fireplace and hatchway foundations must also be marked off and corners squared.

Now, make several batter boards (Figure 10). Eight are needed for a simple square or rectangle, more if there are wings or ells to the foundation. Batter boards are made by connecting a 3-foot 1 x 4 to two 2 x 3 stakes, with a 1 x 2 brace.

Extend the lines of each dimension of the rectangle at least 15 feet, and drive the batter boards into the ground at each 15-foot extension (Figure 10). Now, you can secure surveyor's string to the batter boards so that they cross at each corner. The strings can be adjusted to make the corners square. Saw kerfs (the width of the cut) can be made in the board to secure the string.

Another technique is to make the batter boards in the shape of an L and set them about 4 feet from each corner. However, this is less practical because they're likely to be disturbed by the excavator.

To make sure the strings cross at each corner, drop a plumb bob (weighted string) from the string so that it lines up with each corner. Accuracy is all-important. In fact, this is

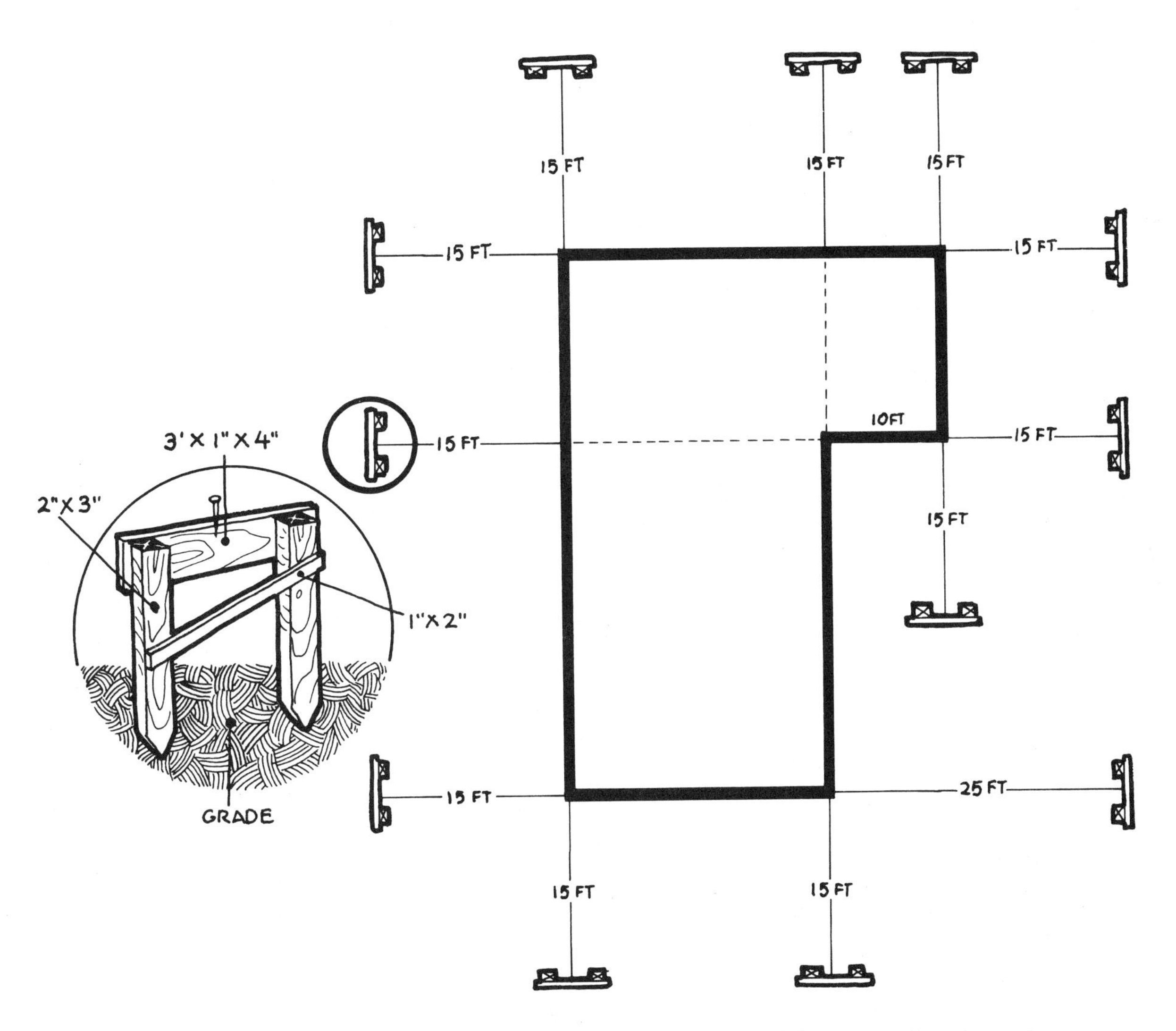

FIGURE 10. Use of batter boards (inset) to allow the stretching of the surveyor's string to form the outline of cellar excavation. Batter boards should be at least 15 feet from the excavation so they will not be disturbed during digging. Batter boards are also used to locate footings and foundation walls.

where accuracy is the most important, because the corners will determine the foundation walls, and if they are off, the whole house will be off. And don't think it hasn't happened.

To determine the depth of the foundation, and also the height, use the highest elevation of the planned building's perimeter as a control point. Then all you have to do is keep those strings level. That's where the string level comes in handy.

You may want to have this done. If not, it is a real challenge to your ingenuity and patience.

The excavation is a must for subcontracting. It isn't too hard to find a good digger who will do the job in a day. He will dig according to your instructions, usually 1 foot outside the outside wall of the foundation. If the earth is pretty solid, the walls can be fairly steep. If not, they will have to be sloped.

While the digger will follow your instructions, keep off his back. An excavator who once did a job for me told me during the work that took three hours, "I could take a day to do this, if you kept telling me what to do."

How deep should he dig? Well, how deep should the footings be? The depth is determined by the height of the foundation walls. Standard is 7 feet 4 inches, from the top of the footings to the bottom of the floor joists, but 8 feet is much better. And because the floor is usually 3½ inches thick, and poured on top of the footings, this height is reduced by that much.

Let's say the land is level. The top of the foundation should be 18 inches to 2 feet above the final grade (earth line) in order to protect against termites (Chapter 25) and to prevent water problems with the wood sheathing and siding. Let's say the exposed foundation must be 2 feet. Then, perhaps the land will be graded another 2 feet above the original ground level and sloped away from the foundation for good drainage. This means that 4 feet of the foundation will be below the original grade. So, the excavation must be 4 feet deep. Of course, this can vary according to how much you want to slope the land down from the foundation. Anyway, the shallower the excavation, the less chance there is for water seepage.

Now, the whole floor will be dug 4 feet deep. Then, a trench is dug for footings, which must be placed on undisturbed earth, to prevent compressing of soft earth and possible collapse of the footings. Or, the whole floor can be dug to the depth of the bottom of the footings, and then gravel put in where the basement slab will go. Six to 18 inches of gravel is not too much. The gravel not only brings the basement slab to the level it belongs at, but provides for drainage under the slab.

Footings are simply a concrete pad onto which the foundations are poured, giving them needed stability. They can be built with board forms or, if the earth is solidly packed, can be poured right into the earth trench. Sometimes, if building codes allow it, and the earth is hardpan or other extremely solid material, the footings can be omitted.

Once the hole is dug, the corner strings, if they've been disturbed or removed, can be checked for accuracy and replaced. Using the plumb bob again, the outside corners of the foundation can be determined.

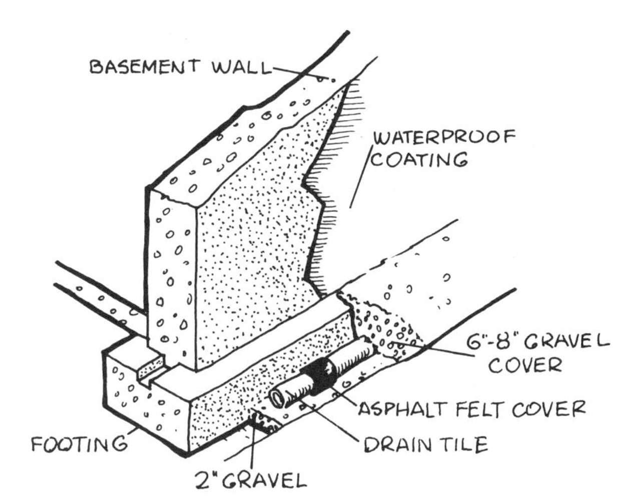

FIGURE 11. Footings and foundation walls of poured concrete. The footing should be as deep as the foundation is wide, and twice as wide as the foundation thickness. The notch, or key, in the footing is made by inserting 2 x 4 when concrete is poured; it prevents the foundation from moving laterally. Drain tile draws underground water away from the foundation and basement.

Enough space must be left in the excavation to permit workmen to install forms for footings and foundation, and to lay drain tile on the outside of the foundation. A good digger will build ramps in the hole to allow concrete and gravel trucks to get as near the foundation as possible. And a good man will come back after foundation forms are installed just to backfill (replace the earth on the outside of the foundation) these ramps.

Communities have specific requirements for the compression strength of concrete (weight directly on the material measured in pounds per square inch). If you order ready-mixed concrete, just specify the strength of the material, according to code requirements.

The ratio of cement to sand, aggregate, and water is what makes it strong. A good, routine type of concrete has a ratio of 1, 2, 3: 1 part cement, 2 parts sand, and 3 parts aggregate. There are many ratios, but it's sufficient to say that the ready-mix people know what they're doing.

Concrete sets more slowly as the temperature goes down, and generally, the slower its cure the stronger it will be. On hot days, when it sets fast, it must be kept wet to cure properly. This means that once it is poured and begins to set (usually in 15 minutes), it must be kept wet by covering it with straw and keeping the straw wet. Or wet down the concrete and cover it with canvas. Concrete usually takes 28 days to cure completely, but it will be hard enough to allow work the next day.

At temperatures between 40 and 50, it will set up and cure nicely by itself. Don't pour concrete in subfreezing weather unless you can apply artificial heat until it sets.

Ready for building. And considering the work so far, it's a good idea to hire a concrete form outfit, which will form footings, pour them, put in reinforcing rods, form foundation, and pour them.

Footings (Figure 11) hold up the entire house, from foundation up. They must be below the frost line, which is the maximum depth to which the soil freezes. This can be as much as 4 or 5 feet in northern parts of the United States. Codes will tell you how deep the footings must be.

Footings are generally the same depth as the thickness of the foundation wall, and should project beyond the foundation on each side by half the thickness of the foundation. If you plan a 10-inch foundation, standard, then the footings should be 10 inches thick and 20 inches wide.

Boards, 2 x 10s, can be used as footing forms. They are secured by stakes inside and out, and must be level both across their breadth and length. If you oil the boards before using, they'll come away from the concrete easily when it has set. Don't use anything lighter in weight.

For added strength, reinforcing rods (rebars) half an inch in diameter can be installed in the footings. They can be supported by small stones so they won't fall to the bottom of the forms while concrete is poured. They also can be bent to turn corners. Rebars are essential where a footing crosses pipe trenches.

It is important to install a 2 x 4 key for poured foundations. This is done by inserting a 2 x 4 into the middle of the footing before the concrete sets (Figure 11), which allows the foundation wall to be keyed into the footing so it will not move sideways.

Stepped footings are built when the land slopes too steeply to have a full level run. Each run of footing must be level. It is important to remember that the height of each step down or up should not be more than the thickness of the adjacent footing.

It's also a good idea to install drain tile outside of and next to the footings, particularly if there might be a drainage problem (Figure 11). Even if there isn't, it's smart to do this because once the excavation is backfilled, the tile is expensive to install.

The drain tile picks up underground water and guides it toward a dry well (if the soil is permeable enough to allow drainage), into a sump to be pumped directly to a storm sewer, if any, or as far away from the foundation as possible. Drain tile is pitched slightly toward the sump, dry well, or sewer. It is made of clay or concrete, 4 inches in diameter and 12 inches long, placed on 2 inches or more of gravel and spaced ⅛ inch apart. The top of the tile joint is covered with roofing felt and the pipe covered with 6 to 8 inches of gravel. You can also use plastic or asphalt and fiber pipe with holes in it, covering the pipe in the same manner as the clay or concrete tile.

Foundation walls are made by pouring concrete into forms, usually made of plywood, fastened to 2 x 4 frame, and reusable (Figure 11). They must be braced. The forms must be both plumb and level, and the level requirement goes two ways: across the width and along the length of the wall. Forms must be built to accommodate pockets for steel and wood beams, the ones that hold up the floor joists. Forms must be built for windows, doors, and such things as fireplaces, hatchways, and wings.

Avoid putting in windows and doors with their tops below the top of the foundation wall; in order for the concrete to span such openings, it must be reinforced with steel rods. If this is necessary, you can install preformed concrete lintels instead of pouring the lintels in place. When the window and door openings are level with the top of the foundation wall, the wooden sill will span them, a perfectly good construction technique.

Form men work miracles. In today's modern concrete construction, forms turn around corners, can be installed at a slant, and can be surfaced so that the concrete will take their form, part of the design.

But you will be happy with what a crew can do. It's exciting when the concrete trucks arrive, huge monsters disgorging their gray, liquid mass that seems so unlikely to hold up your house. The concrete men are very helpful, but you must be prepared, if the form men don't do it, to puddle and vibrate the concrete as it is poured so that it will be as compact as possible and will flow into every nook and cranny.

Concrete work is hot, hard labor. If you insist on helping out, fine and dandy, but you're likely to get your feet into it, and get it on your hands. It is corrosive and abrasive; get your hands wet with concrete too much and you'll find the old pinkies getting mighty sore. And your hands will feel slippery. Don't worry, that's simply the first layer of skin wearing off. So wear heavy gloves and shoes and clothes, or try to keep the concrete in the forms and off of you.

The form men will make sure the foundation wall stops at the right level. Most of the time it will stop below the top of the forms; it isn't a matter of just leveling the concrete at the top. A proper job will make the top of the foundation walls pretty smooth, so the sill will set snugly.

Concrete must be poured in one continuous operation, or at least should be. If you fill half the foundation and knock off for a day and then continue, you'll have a horizontal seam that could leak and, if above the grade line, be unsightly.

At the top of the foundation wall, anchor bolts are inserted in the fresh concrete. Anchor bolts are ½-inch steel bolts curved at one end and threaded on the other to take a washer and nut. Embed the bolts, curved end down, in fresh concrete, at 4- to 8-foot centers, with enough threaded end exposed so that it will go through the sill to allow the nut to snug the sill onto the foundation. Sill sealer, a strip of semirigid fiber glass, goes on top of the foundation to fill any irregularities between the foundation and the sill.

A termite shield is a good idea to install in most parts of the country, because termites can be troublesome in all but the extreme northern areas of the nation. The shield is a metal strip installed on top of the foundation to form sort of a hood over it in the outside, preventing termites from building their mud tunnels up to the foundation and into the wood. Termite details can be found in Chapter 25.

In most houses, the width of the foundation is far more than one floor joist will properly span. So a large beam is set in the foundation walls to cut this span in half. The beam, even a good steel one, must be supported by columns, so footings must be poured at the same time the foundation is poured, to support these columns. They should not be supported by the concrete basement floor, which is poured later. Post or column footings are usually 12 inches deep and 2 feet square. In rare cases the footing for a column is 30 inches square, but that size is generally not necessary.

When steel columns are used (they come filled with concrete), they must have top and bottom plates to be bolted onto the beam and the concrete footing. The bolts are set in the concrete as it is poured. When wood posts are used (less satisfactory but adequate, particularly when treated with wood preservative), a steel pin is set in the concrete as it is poured and the post inserted over the pin. Sometimes a pedestal of concrete is used.

Once the posts are in place holding up the beam the basement floor can be poured. The actual pouring may be one of the last things to do in the project. (See Chapter 27.)

Foundation walls can be waterproofed by applying a layer of hot or cold asphalt (hot is better) from the footing to the grade level. Better yet is a layer of hot tar, then overlapped 15-pound roofing felt, another layer of tar, a second layer of overlapped felt, and a third coat of tar. Overlap the felt by half its width.

If you're hipped on doing some foundation work yourself, you can try concrete blocks, put together with mortar. Lots of luck. Concrete blocks weigh nearly 30 pounds each, and under general masons' rules, laying them is a two-man job. Blocks are a nominal 16 inches long, 8 inches wide and 8 inches high. Subtract ⅜ inch from all three of these dimensions and you'll get the actual size. A ⅜-inch cement mortar joint brings the block to full size. The best way to lay blocks is in a common bond, with the second course (layer) offset by half the length of the block in the course below. Some codes require 12-inch-wide blocks for foundations.

Blocks and bricks are laid in cement mortar, made of cement, hydrated lime, and sand. Don't use beach sand near the ocean; it contains salt, an enemy of concrete and mortar. For regular service, use 1 part Portland cement, 1 to 1½ parts lime, and 4 to 6 parts sand. Or, use 1 part masonry or mortar cement (with the lime already in it) and 2 to 3 parts sand. You can also buy ready-made mortar mix — all you have to do is add water — but it is very expensive. Mortar should be on the dry side; too much water will make it soupy, sloppy, and weak. Dryish mortar tends to keep off the face of the block. It should be pliable enough to adhere to the side of the blocks without falling off.

Blocks must be laid in a bed of mortar, with each succeeding block mortared where it touches its neighbor. A string should be stretched at the right height so the blocks line up properly and level. Check them regularly with a 4-foot spirit level, both for vertical plumb and horizontal level.

The tops of a block wall must be capped with a 4-inch-thick solid cap, or concrete reinforced with wire mesh. Anchor bolts are inserted on 4- to 8-foot centers between the caps and penetrate the holes in the blocks. Fill each hole containing an anchor bolt with concrete or mortar. Fill the two top courses. Use plenty of stones in the hole; it will save on mortar. A block wall must be protected with a coating of mortar, and then a waterproof membrane coating like that described for concrete walls.

If you decide that a basement is not for you, or ground or water conditions prohibit it, then consider a crawl space.

Footings for a crawl space foundation must be below the frost line and the foundation extend the same distance above grade as for a house with a full basement. Obviously, there is no need to excavate much (if at all) inside the crawl space, except enough so the bottoms of the floor joists are at least 3 feet above ground level.

Construction is similar to that for full basements except there are no windows and the walls can be thinner. Generally, a 6-inch poured-concrete wall is sufficient, but 8 inches is better, both with comparably sized footings: 8 to 12 inches deep and twice the width of the walls. Concrete block can be 8- or 12-inch.

Concrete or block piers take the place of columns to hold up the large beams. A product called Sonatube makes excellent concrete piers. It is a heavy fiberboard cylinder set on the proper-sized footing and filled with concrete. The fiberboard is pulled or peeled off, and presto! You have a round pier. The tubes come in several sizes. Concrete block piers also can be built, on the proper footing. It is best to pour or build the piers before the large beams are installed, to ease the operation. It's only a matter of making sure the tops of the piers are at the proper level, and if they are short, shims (wedges) can be inserted to snug the pier and beam together.

A crawl space should be ventilated, so pockets should be installed in the forms for the concrete wall. With concrete block construction, a block can be left out at the top of the wall. The hole is covered with screened, louvered ventilators. The ventilators should be kept open all year around, although some experts recommend they be closed in winter. The idea of ventilators is to allow air circulation to prevent buildup of moisture, which can cause fungus growth and decay of the floor joists.

The earth floor of the crawl space should be covered with 4- to 6-mill polyethylene or roll roofing to prevent moisture from coming up through the earth. The polyethylene should be lapped by 2 or 3 feet and the roll roofing lapped 6 inches and secured with roofing cement.

Although this book is basically about how to build a frame house, if you plan brick veneer walls or even a partial wall such as the front wall up to the second floor, the brick must be laid on a proper base.

Full brick walls are two bricks thick and stand by themselves. Brick veneer is one brick thick and is laid up against a wooden wall of studs and sheathing with air space between. For brick veneer, the top of the foundation should be offset about 5 inches, forming a shelf on which the first course is laid. The shelf is 3 courses below the top of the foundation, so that flashing (sheet metal) can be applied to the third course connecting the sheathing and brick, allowing drainage. Metal ties must be fastened to the sheathing or studs and mortared between the bricks, literally to tie the brick to the sheathing. The ties should be 16 inches apart vertically (every fifth course) and 32 inches apart horizontally (every four bricks).

The offset on top of the foundation is required by some community building codes. If the codes allow it, you could start the brick veneer and the wood wall at the same level on top of the foundation. This would not be as good for drainage, however.

Sheathing paper (or slater's or roofing felt — light and heavy tar paper) should be installed on the sheathing whether the wall is of brick veneer or covered with wood siding.

Many builders do not build basements, using instead the crawl space or a slab of concrete on the ground. The slab on the ground, or slab-on-grade (Figure 12), is more popular in warm climates, but a slab or crawl space is a powerful argument for economy, particularly when there is trouble with ground water or water drainage. We re-

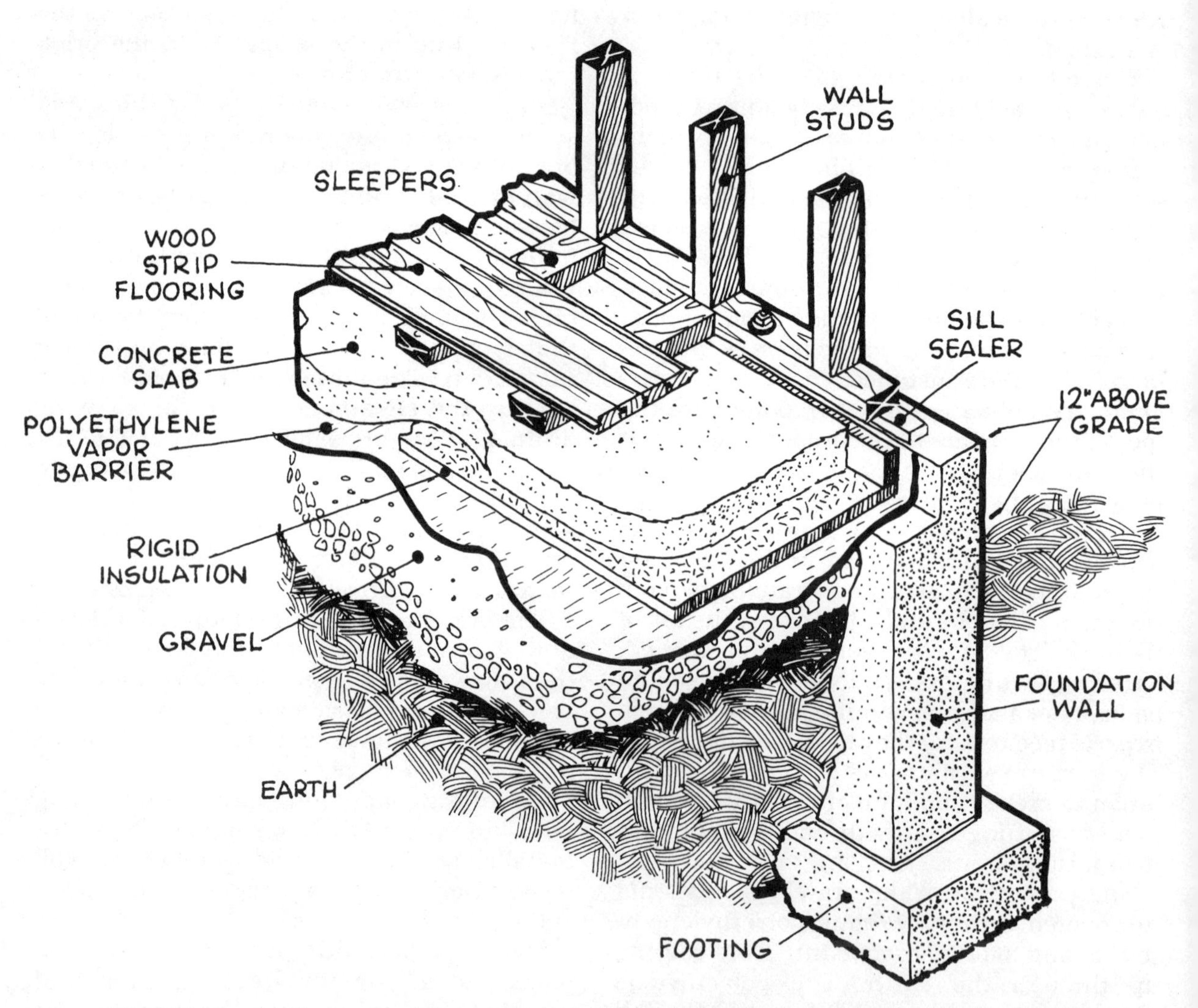

FIGURE 12. Insulating for slab on ground. Polyethylene membrane should be used under all slabs. Sill sealer is made of glass fiber. Sleepers are one technique for providing wood floor on concrete slab.

peat that if the contour of your property lends itself to a "walk-in" basement, and drainage problems are absent or solvable, then the basement is a great idea because it will provide good live-in space.

But if there is a water or drainage problem, here's how to build a slab-on-grade. It needs footings and foundations just as a crawl space or basement does, because if you don't get below the frost line, the slab will heave and you'll be living in a crooked house. Maybe that's where the nursery rhyme about a crooked man came from.

In warm climates, a monolithic slab is poured, meaning that the slab and foundation and footings are one piece (literally, one stone). This is done by excavating a perimeter ditch and pouring the foundation (using forms on the outside perimeter) or using the excavation itself as forms, and pouring the footings, foundation, and slab all at once. Reinforcing rods and/or 6-inch wire mesh (the opening of the mesh is 6 inches square) should be used where the slab meets the foundation and all along the slab.

Basically, the slabs should be high enough above grade to allow for the ground to slope away from the foundation. Eight inches is usually minimum, but 12 inches is better. The slab should be 4 inches thick, laid over 6 to 12 inches of gravel or crushed stone, and reinforced with ½-inch rebars or 6-inch wire mesh.

In cold areas, the slab should be separate from the foundation. And between the slab and the foundation, install 1-inch rigid insulation going down to the bottom of the foundation or turning horizontally to extend completely under the slab. The insulation must be above a vapor barrier (see page 197). Insulation can be glass fiber (more concentrated and more rigid than the familiar glass fiber "cotton candy" insulation), plastic (polystyrene, polyurethane, urea formaldehyde and others), but it must be water-resistant and able to stand up against the weight of concrete. Another way to keep the slab warm (a common complaint of concrete slabs is that they are cold) is to install heating ducts in the perimeter or heating pipes in the entire slab. This is good heat, but if anything goes wrong with those ducts or pipes, it's a major job to fix.

All things considered, a crawl space or basement is better than a slab in cold climates. The country is generally considered pretty cold in the winter north of a line that cuts the nation into north and south halves. The northern plains are particularly cold, and the Northwest is particularly warm in coastal regions. Local building codes and authorities can give you a good idea of what is best.

Even if you've solved the problem of cold concrete floors on a slab, they can still be plenty hard, and the hardness is unrelieved by wall-to-wall carpeting or resilient (vinyl and vinyl asbestos) or wood tiles.

But there is one solution. Put down 2 x 4 sleepers on their wide side, on 12-inch centers. Because you have installed a vapor barrier beneath the slab, you shouldn't have to worry about moisture. The sleepers must be anchored: by concrete nails, a nail gun (nails driven by a .22-caliber blank, very dangerous for the layman), or adhesive. On top of this apply ⅝- or ¾-inch plywood, and on top of this carpeting, wood, or resilient tiles.

You can even put 1½ inches of insulation in the air space between the sleepers. The 12-inch centered sleepers make it possible for only one layer of plywood; otherwise two layers would be needed.

Another reason for the sleepers and plywood floor is that they reduce or prevent condensation of house moisture on the cold concrete floor. The heated floor would solve this problem in winter, but bare concrete (even covered with resilient tiles) could cause problems in off-seasons.

The concrete slab can be finished off in two ways (Figure 12). If you plan to cover it with resilient or wood tile, you need a smooth surface, made by smoothing with a metal trowel. If you're going to cover it with sleepers and plywood, it can be troweled with a wood float, making a somewhat rougher surface.

chapter 8

The End of the Beginning

Sills, Floor Joists, and Floor

So far, you've probably stood around watching the pros dig the foundation and install footings and concrete walls.

Now it's your chance to shine, because it's the start of framing, working with wood, which is probably an easier thing for the amateur to do (Figure 13). It's the time for the builder (you) to get a companion, and if you have to hire a carpenter to work with, fine, because he'll be able to help you more than a fellow amateur could. At any rate, two men are essential to wrestle with sills, 2 x 10 joists, and other long, heavy material.

First wood to go on the foundation is the sill, or sill plate (Figure 14). This can be a 2 x 6 flat, or two 2 x 6s nailed together. Some codes require the sill to be a solid 4 x 6, shiplapped at the corners. The sill should be treated with a wood preservative and laid on a fiber-glass sill sealer.

Holes are drilled through the sill to coincide with the anchor bolts in the foundation. The sill is slipped over these bolts and secured with washers and nuts. Don't snug them down too much; they should be straight and level. Too much snugging will make them follow the contour of the foundation, which is almost never straight and even. The sill sealer will make up for variations in the top of the foundation. The sill is placed ¾ inch in from the outside edge of the foundation wall to allow for sheathing to butt down against the foundation. The sill plate is secured in the same manner on a crawl space foundation. We are describing the construction of a platform framing system. Platform construction consists of floor platforms, on which walls are built, which in turn support the second floor.

Another type of construction is called balloon framing, which is slightly stronger, especially for a two-story house, but is more complex to build, particularly for the amateur. In balloon framing, floors are virtually hung on the vertical studs, which go the entire length of two full stories. Studs this length are not only expensive but may be troublesome, because the longer the studs, the less straight they might be. Balloon framing was invented in the nineteenth century, taking the place of post and beam construction, which was just about the only way the Colonials and early Americans knew how to build. It was called balloon by a skeptic who said it wasn't any stronger than a balloon. He was mistaken.

Because most houses are too wide or deep for floor joists to span in one length, a large

This is a Cape Cod type house, with roof rafters running up from the top of the first floor.

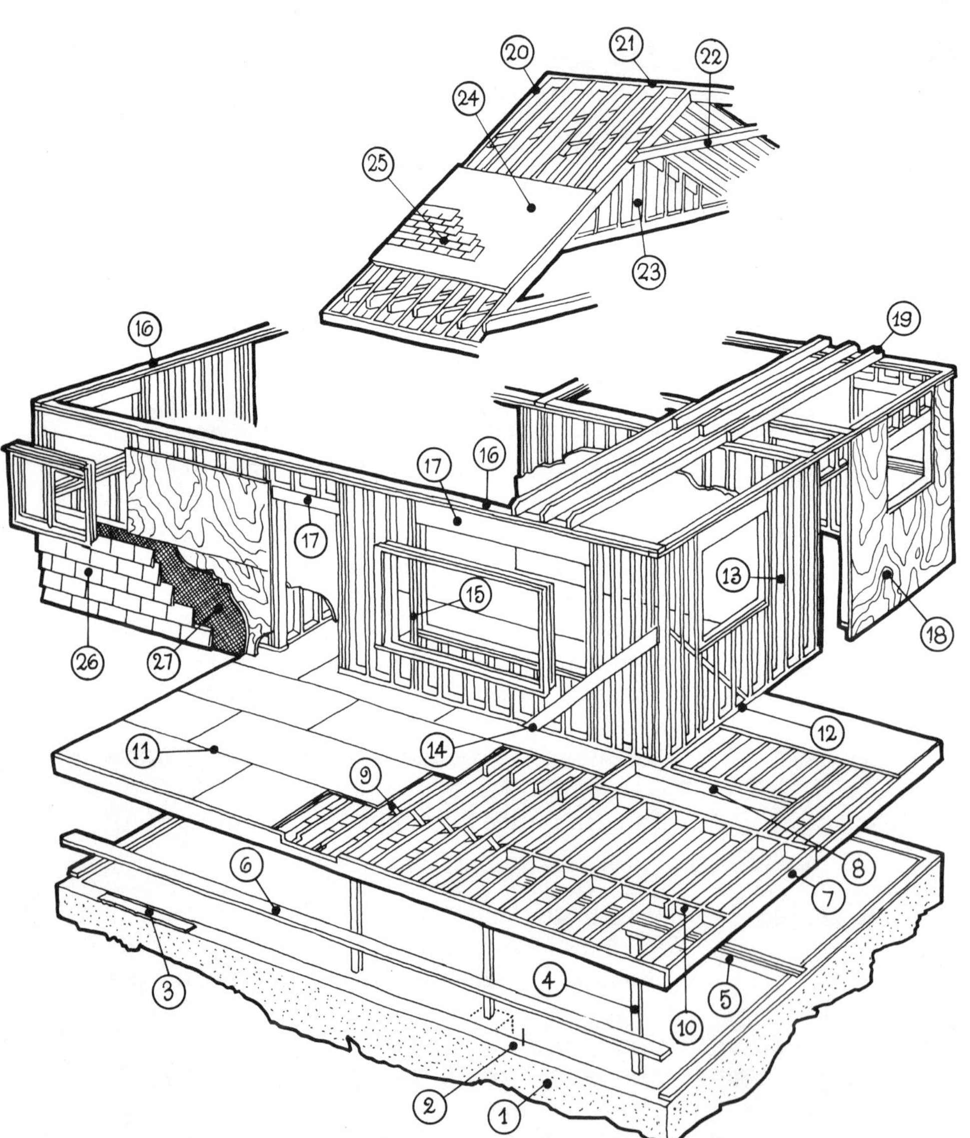

Key:

1. *Foundation*
2. *Anchor bolt (every 4 to 8 feet)*
3. *Sill sealer (fiber glass)*
4. *Post to support center beam*
5. *Center beam (three wood or steel 2 x 12s spiked together*
6. *Sill (4 x 6)*
7. *Floor joists (2 x 10)*
8. *Doubled header joist for openings in floor*
9. *Cross bridging*
10. *Solid bridging*
11. *Floor sheathing (plywood or matched boards)*
12. *Floor (sole) plate (2 x 4)*
13. *Studs (2 x 4)*
14. *Corner brace*
15. *Jack studs (doubled or tripled at door and window openings)*
16. *Top plate (2 x 4 doubled)*
17. *Header (2 x 8 doubled or 2 x 4 doubled)*
18. *Wall sheathing (plywood or matched boards)*
19. *Ceiling joists (2 x 8 or 2 x 10)*
20. *Rafters (2 x 8)*
21. *Ridgeboard (2 x 10 or 1 x 10)*
22. *Collar beam (1 x 6 or 2 x 6)*
23. *Gable studs (2 x 4)*
24. *Roof sheathing (plywood or matched boards)*
25. *Roofing shingles*
26. *Wall siding (shingles or clapboards)*
27. *Sheathing paper (building paper or roofing felt)*

FIGURE 13. *Exploded view of framing, from ground floor to ridge, showing sheathing, siding, and roofing techniques.*

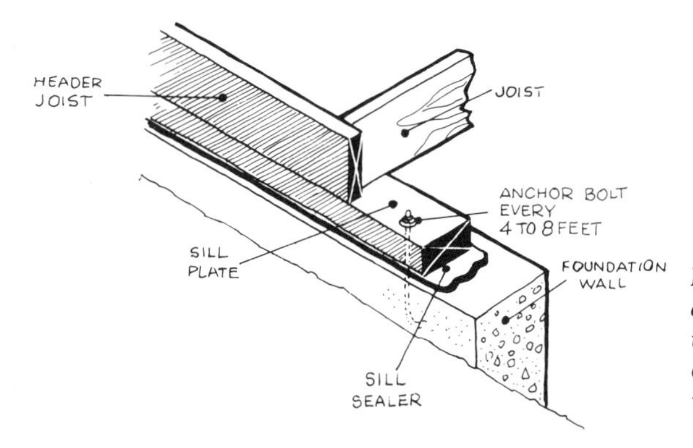

FIGURE 14. The sill, a 4 x 6, sets on a sill sealer on foundation wall, and is set ¾ inch in front of the foundation edge to allow the sheathing to come flush with the concrete. Anchor bolts are set when concrete is poured.

beam splits the width in half. Say a house is 25 feet wide; the beam would cut this width to 12½ feet. A beam can be wood (a 6 x 10 or three 2 x 10s spiked together) or a steel I beam. Steel is a little more expensive but is recommended.

If you use a wood beam, you can spike three 2 x 10s together with 20d nails, two near each end and the others staggered on 16-inch centers. If the nails go through the material, the points should be clinched (bent over). This makes a very strong connection; unless you straighten out that clinched nail, it's likely that you'd have to rip apart the wood to get the pieces apart.

The beam sets into pockets in the foundation with a minimum of 4 to 6 inches bearing on the foundation. Treat the beams with preservatives where they set in the pockets and, to prevent moisture from building up and causing rot, leave ½ inch of air space at the sides and end. An aluminum or galvanized steel termite shield can be wrapped around the end as well.

The top of the beam should be even with the top of the sill plate, so that floor joists can be set on the sill and the beam. Posts (wood- or concrete-filled steel) are set in place on their footings before the beam is set. A wood post is drilled to fit over a pin inserted in the footing while it is poured. A steel post has its base secured by bolts in the footing. The height of the posts is determined by measuring from the top of the sill vertically to the top of the footing, subtracting the depth of the beam. Once the posts are in place, the beam is set in the foundation pockets and on the posts, and the posts are secured, wood with angle irons held by lag screws and steel with a top plate lagged onto the beam.

A steel beam is set the same way, except a 2 x 6 or 2 x 8 (depending on the width of the beam) is bolted on its top to act as a nailer (nailing surface) for the floor joists. It is best to use steel columns with a steel beam. The top plate for the post is bolted onto the beam. If you have specifications on where the posts will go, you can have the bolt holes predrilled when ordering the beam. Otherwise they have to be drilled on the job, a considerable undertaking.

Sometimes the floor joists are butted

against the beam, or to put it another way, hung on the beam. You can use a 2 x 4 ledger nailed directly on the bottom edge of a wood beam. Use two 16d nails at each joist. Or you can use joist hangers, marvelous items that take the place of a ledger. With a steel beam, the joist can be set on a ledger bolted to the bottom edge of the beam (setting on the flange) or directly on the flange. For the amateur, however, setting the joists directly on the sill plate and beam is the best and easiest technique.

Floor joists should be 2 x 8s for spans less than 10 feet; 2 x 10s for 10 to 16 feet; and 2 x 12s for more than 16 feet. Anything less than these sizes will cause sagging and bouncy floors.

Floor joists may vary a little in width, so they must be placed first without nailing. Use a very long straight edge (a straight 2 x 8 is good) to make sure each joist is level with its neighbor. If a joist is high, it will have to be notched where it meets sill and beam. If it is low, it must be shimmed. Use shingles for shims; they are tapered and can be inserted to shim 1⁄16 to 1⁄4 inch; two can shim up even more. A discrepancy in joist level of 1⁄8 inch is allowed.

Also, place the joists with their crown edge up. If you sight down a joist, you'll notice a slight bend. You also may be able to tell by looking at the grain: if the grain or joist crown forms a raised curve, put this edge up. The crown will level off with the weight of the subfloor and house.

Joists are placed on 16-inch centers. In fact, all joists, rafters, and studs are on 16-inch centers in standard construction. And this means 16 inches from the center of one joist to the center of its neighbor. The first joist is a stringer joist, flush with the sill plate. Because the width of the house is probably more than the maximum length of the joist you can buy, butt the stringer joists together end to end and splice them with a nominal 2-inch plank, using 12d or 16d nails and clinching them if they extend through the joist.

Then line up succeeding joists, overlapping them by a minimum of 4 inches (12 inches is better) over the center supporting beam. The first pair of overlapping joists parallel to the stringer joist will be on each side of a mark 16 inches in from the stringer joist. Then, each succeeding joist can be 16 inches on center from its neighbor, except where openings are necessary for stairways, chimneys, and fireplaces.

When you come to the other end of the foundation (the opposite stringer joist from the one you began with), any space left between the last joist and the stringer joist should be less than 16 inches on center. If it is 24 inches, for instance, add another joist on a 16-inch center, leaving a final gap of about 8 inches. In other words, the final gap of any joist or stud system should be less than 16 inches.

To measure 16-inch centers easily, measure from one side of the first joist to the same side of the next joist. Draw a line with a square at this mark. Then, when the joist is placed, its center will line up automatically 16 inches from its neighbor. Try it, it works. Or, you can cut a spacing block 14½ inches long and use this to measure the gap between the joists. Joists set on 16-inch centers are 14½ inches from edge to edge. Not only will this block space the joists automatically, it will also help hold them in place when toenailing (see below) is necessary.

Why all the fuss about 16-inch centers? Something had to be standard, and when the standard was set, materials were strongest on this measure. Paneling and wall materials like plasterboard and plywood were dimensioned to accommodate the 16-inch standard. Four-foot-wide material will span 4 joists or studs.

Currently there are proposals to establish a standard of 24-inch centers, to save on dimension lumber like joists, studs, and rafters. In fact, some codes and specifications allow rafters (roof beams) to be 24 inches on center. This depends on what is used for rafters, roof sheathing, and the snow load in winter. At any rate, the 24-inch system is not too bad, if you use thick enough material. The new adhesives used in some commercial construction help make the 24-inch system stronger than with other kinds of fasteners but most codes still do not allow the 24-inch system.

This Mod 24 system may be the building technique of the future, for two reasons. One, it saves on building materials (fewer joists, studs, and rafters are used) and therefore cost (about $200 a house). Second, it would work well with the increased insulation that is being recommended by insulation engineers.

These engineers are now figuring that 5½ inches of fiber glass insulation in the walls, instead of the standard 3½ inches, is important to save fuel. (See Chapter 18.) Well, we all want to save fuel. But such space for insulation would require outside walls to be 5½ inches thick. Such walls would have to be made of 2 x 6 studs. The increased cost of using 2 x 6 studs instead of 2 x 4s would be offset by setting them on 24-inch centers. Interior partitions and bearing walls would still be made with standard 2 x 4 studs.

It's certainly something to think about. You can use 2 x 6 studs with a standard plan, but an extra-thick wall would use up a little more interior space, and you must anticipate that window and door jambs would have to be wider to accommodate the thicker wall.

But think of the fuel you would save.

The length of your joists must be 3 inches short of the width of the house from sill to sill. This will allow for the two header joists, each nailed against the ends of the joists on each long side of the house. So now the header joists line up with the sill.

Once the joists are lined up and level, toenail them to the sill and center beam with two 10d nails. Toenailing simply means driving a nail at an angle into the material being held and then into the holding material. Toenailing is done where it is impossible to face-nail directly through the material being held.

Then, nail the header joists to each stringer joist with three 16d nails. A stringer joist is a border joist of a floor frame, parallel to intermediate joists. The header and stringer joists are nailed to the sill with 10d nails spaced 16 inches. Finally, the joists are nailed into

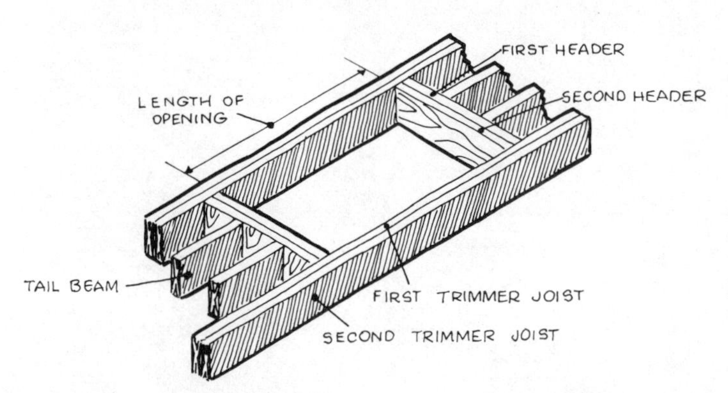

FIGURE 15. Doubled joists and headers are specified for stair and chimney openings in the floor. Be sure to install the material in proper order so you will be able to face-nail directly through each layer of material into its neighbor.

each other where they overlap over the center beam with three 12d nails. Clinch any points sticking through the wood.

Where openings in the floor are required, for stairwells, chimneys, and fireplaces, the joists are doubled (Figure 15). Cross-joists, called headers, are placed where the joists stop, and are doubled. Generally, nailing is done first with single joists and headers, then the doubled joists and headers are nailed to crosspieces and each other with 16d nails at each joint. Joists are also doubled under room partitions.

Bridging between joists is recommended midway in the span from sill to center beam. Some experts maintain that this does not add to the strength of the floor once the subfloor is installed, but we think it does help to prevent warping and the tendency of joists to lean; that is, not sit squarely on sill and center beam. Cross-bridging is comprised of crossed 1 x 3 boards connecting each joist, and solid bridging uses short lengths of planks, the same size as the joists, connecting each joist. Nail the cross-bridging 1 x 3s with two 8d nails. Nail the solid bridging with two 12d or 16d nails at each end. You can face-nail each end of the solid bridging by staggering the position of the planks.

If you plan bay windows or any type of projection of a floor beyond a wall (garrison type houses have the entire second floor projecting over the front wall), you'll have to learn the technique of cantilevering.

To build a projection, where joists are perpendicular to the header joist, simply extend the regular joists where needed, and double the joists that form the borders of the projection (Figure 16). A header joist is nailed to the extended joists, called lookout joists, which look out over the wall. Technically, they are cantilevered. (That's one thing about construction: the terms are often quaint, but very descriptive and quite constructive.) Filler joists are placed where the header joist would have gone if there were no projection. This fills the projection where it crosses the sill.

Where the projection extends beyond the stringer joist, that is, where joists are parallel to the stringer joist, install lookout joists or stringers from the third joist in from the stringer joist, so that the lookout joists extend 32 inches in from the sill, and 12 to 24 inches out from the sill (Figure 17). The joist that supports the lookout joists is doubled; so are the lookout joists on the edges of the projection. A header is nailed to the lookout joists and filler joists are set between the joists over the sill. Joist hangers at the joints can strengthen this structure. Wherever

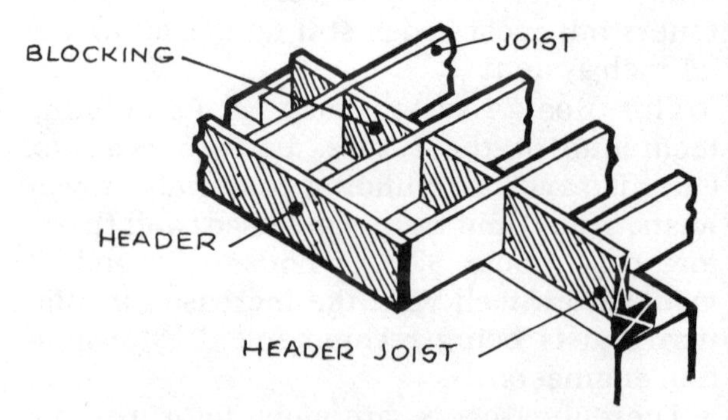

FIGURE 16. First-floor overhang for bow window where joists are perpendicular to header joist.

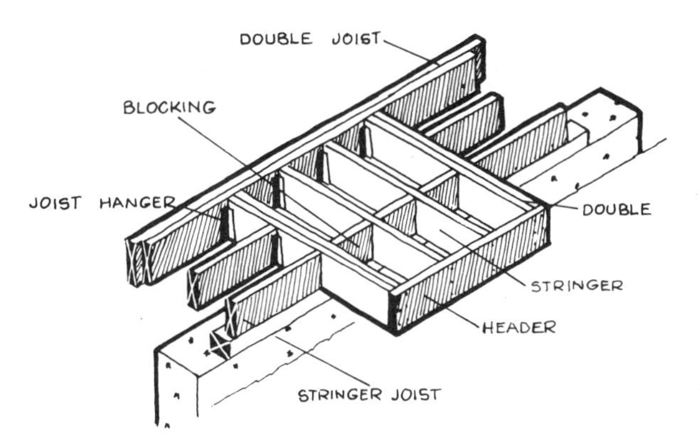

FIGURE 17. First-floor overhang for bow window where joists are parallel to stringer joist.

there are projections, the main wall (not the projection wall) has a header over the opening to support second-floor joists or rafters.

Subflooring is next. It can be boards or plywood, both of which are good when a top floor of oak or similar strip or plank flooring is installed. Either is also good when an underlayment is used to support carpeting or linoleum or if resilient or parquet wood tile is used.

Boards can be square-edged or tongued-and-grooved (T&G), a nominal 6 to 8 inches wide. Current usage recommends they be laid diagonally, at a 45-degree angle from the perimeter. This allows a strip flooring to be laid at right angles to the joists, with a nail in each joist. If subfloor boards are nailed at right angles to the joists, the top floor must be laid parallel to the joists, allowing nailing only in the subfloor, an inadequate procedure. If the floors are laid parallel to each other, cupping or humping of the top floorboards can occur.

Subfloor boards are usually made of spruce, an inexpensive, strong lumber. Nail with two 8d or 10d nails; three if the boards are a nominal 8 inches wide. Where you must butt end boards, butt them over a joist. Generally, lay the boards with their ends slightly overlapping the perimeter. Then cut them with a rotary saw, even with the stringer and header joists.

When using plywood for a subfloor, use ⅝-inch plywood with exterior glue (waterproof or water-resistant), usually a rough quality called plyscore. Install 4 x 8–foot sheets with the grain at right angles to the joists. The joints should be staggered. Nail with 8d nails or ring-shanked underlayment nails. The latter have excellent holding power.

Space the nails 6 inches apart along edges and 10 inches on intermediate joists; that is, where joists lie under the inner areas of the plywood sheets. Joints should be spaced 1/16 to ⅛ inch apart to allow for expansion.

Installing linoleum, tiles, or wood parquet tiles over a layer of plywood is not recommended. If you go ahead and do it anyway, the plywood should be 1 inch thick (full dimension) or even 5/4 (1¼ inches) thick, and tongued and grooved at all joints. You might find this quite expensive, and the face side of the plywood must be smooth enough so defects will not show through the linoleum or tile. Besides, single-thickness floors always are unduly bouncy.

Underlayment for linoleum, tile, and carpeting can be ½-inch plywood or ½-inch particleboard. It is not installed until much of the interior work is done.

chapter 9

Like a Barn Raising

Wall Framing

Wall framing is among the most satisfying parts of building a house, because it transforms the house from a mere platform above ground to a real shelter. The more people you get to help you, the faster the tedious part will be over. And because framed walls are built on the floor and put up in sections, it's a good time for a big party afterward, a little like an old-fashioned barn raising.

Framing uses 2 x 4 vertical studs, set on 16-inch centers. The studs are set on a 2 x 4 called a sole or floor plate, and are topped with a double top plate. The only variations on this technique are at the corners, at door and window openings, and where interior walls attach to outside walls (Figure 18).

Most 2 x 4s are made of fir or hemlock, and under current specifications, a combination of fir and hemlock is allowed in a batch of dimension lumber. Depending on the part of the country you live in, dimension lumber can be southern pine, spruce, or white fir, as well as fir or hemlock. All are dependable for framing.

How high should the walls be? Standard height is 8 feet, from subfloor to the top of the doubled top plate. Because there are three 2 x 4s built into the wall (one floor plate and two top plates) that total 4½ inches, take this off the dimension when you cut the 2 x 4 studs, which then will be 7 feet 7½ inches. So the final height of the walls from finished floor (¾ inch) to finished ceiling (1¼ inches) will be 7 feet 10 inches (2 inches from 8 feet). That ceiling seems to be a bit thick, but this will be explained when ceilings are covered (See page 143).

This ceiling height allows the use of vertical 4 x 8–foot sheets of plasterboard or paneling for inside walls, or two 4 x 8 sheets of plasterboard set horizontally. At any rate, the house plans will specify the height of the ceiling.

Cut your studs to the desired length. Lay the floor plate and one top plate flat on the floor, using blocks or an extra 2 x 4 nailed temporarily to the floor to act as a cleat. Then, using a tape rule that is marked in 16-inch increments, mark on both plates the 16-inch centers where the studs will go.

Because the plates will not be long enough for the length or width of the house, cut and lay the shorter pieces out so they'll meet at a floor joist. Short plates also should butt at a stud.

Once the plates are marked for stud position, cut a 2 x 4 14½ inches long and use it as a stud spacer as you nail the studs. Use two

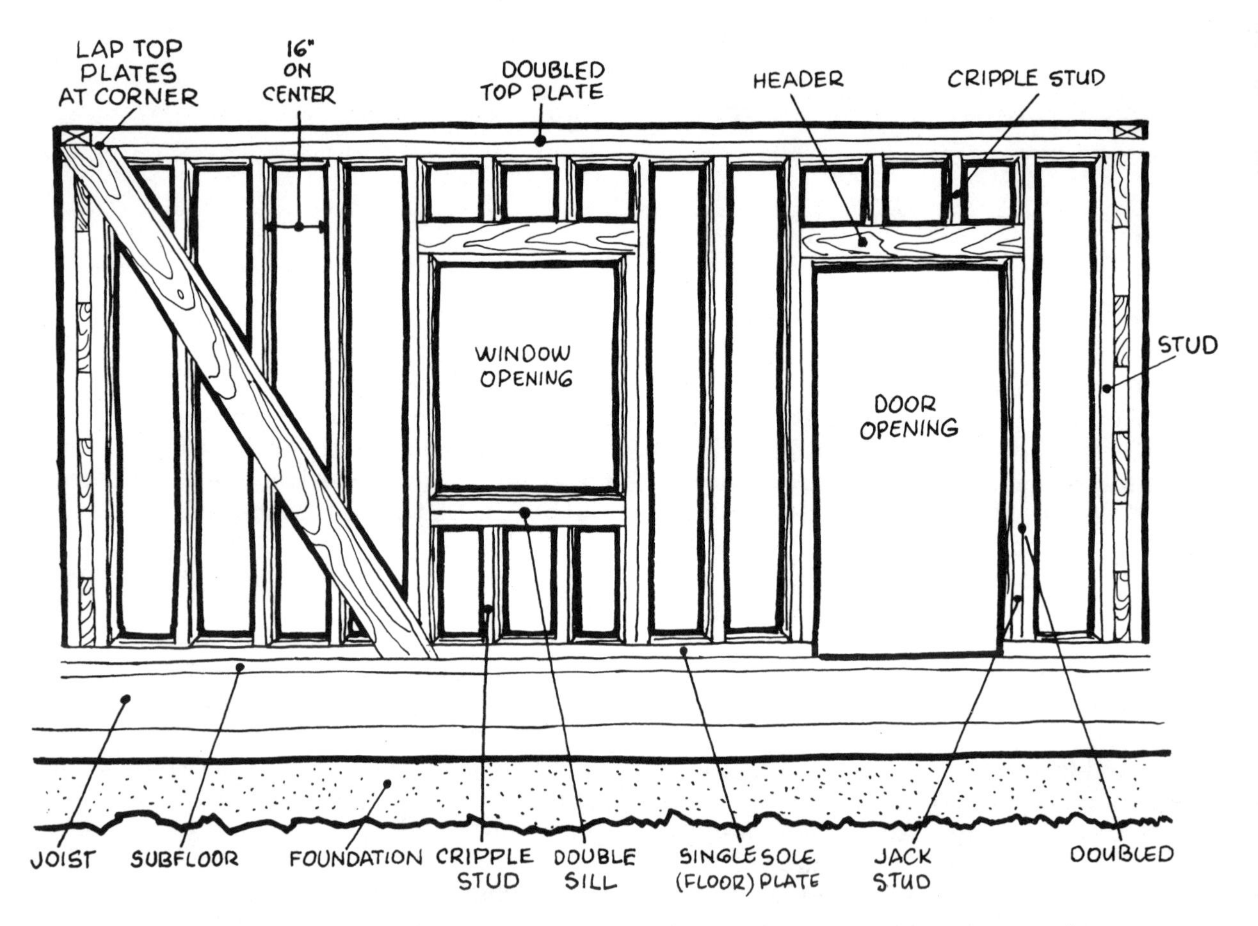

FIGURE 18. Frame wall, showing double studs at door and window openings, corner stud treatment, and a let-in brace (diagonal piece at left) for rigidity. Studs are notched to receive the brace so the face of studs are even with the face of brace.

16d nails to nail the plate onto each stud. Do top and floor plates separately.

Corners are special, because you have to have an inside surface on both walls of each corner onto which to attach the inner walls. So, one wall at each corner must be reinforced. You can do it two ways: Two 2 x 4s separated by one-foot spacer blocks, nailed with 16d nails, or a 2 x 4 stud set at right angles to the end stud on the inside of the plate, also secured with 16d nails. Still another way in quality construction (and sometimes required by code) is to use a 4 x 6 corner post (Figure 19).

The stud wall would be very simple if it had no windows and doors. But we have to make provisions for them, and there are standard techniques employed for the rough openings.

Because windows and doors today come "set up," that is, already in their frames, needing only to be inserted in the openings, you must know the size of the windows before making the rough opening, which must be about ¼ inch wider at each side and top to allow for leveling and plumbing the windows and doors (Figure 18). Actually, the window and door manufacturer will specify the rough opening for its units. The plans also will specify.

Most doors are 6 feet 8 inches (80 inches) high. Some are 78 inches high. The exterior front door is usually specified 36 inches wide, the back door 32 and the interior doors

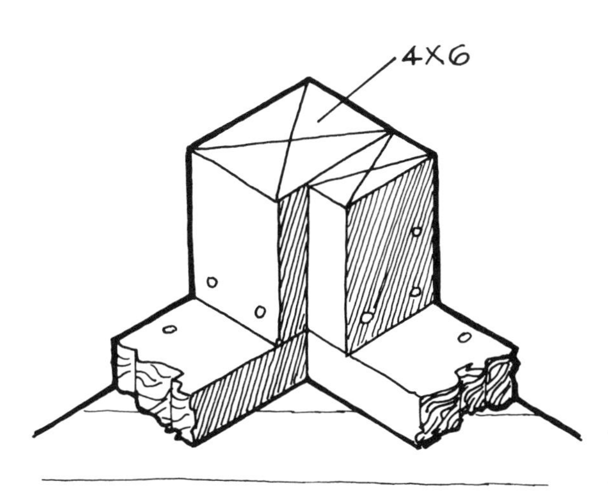

FIGURE 19. Corner stud treatment, using a 4 x 6 to allow a nailing surface on both walls. Or you can use two 2 x 4s separated by 2 x 4 blocks.

30 to 32 inches. Setup doors and windows include side and head (top) jambs and exterior casing.

Here is how to build a rough opening for windows and doors (Figure 18). Determine width of rough opening. Add 3 inches and set full-length studs at these points. Now, cut two 2 x 4s to the height of the rough opening, and nail one next to each of the 2 x 4s framing the opening. Use 16d nails. These inner studs support the header, which can be two 2 x 6s nailed together with 16d nails. Because two 2 x 6s measure only a shade over 3 inches when butted together, which is less than the 2 x 4s' 3½-inch width, insert a spacer of ⅜-inch plasterboard or plywood to bring it to the thickness equal to the width of the 2 x 4.

The two full-length studs are nailed to the header with 16d nails. The header is connected from its top to the top plate by short 2 x 4s, face-nailed through the top plate with 16d nails (shorter ones if the 2 x 4s are short) and toenailed to the header with 10d nails.

In current construction, the tops of all doors and windows are at the same level, so all headers can be of the same dimension, and the short studs connecting the headers with the top plate also can be the same. Sometimes, with different ceiling heights, the header goes snug against the bottom of the top plate, eliminating the need for short connecting studs. Headers are of the same size except for extra-wide door or window spans.

Now, windows are done the same way, only different. The doubled studs are set up in the same manner, and so is the header, but if the window is 3 to 5 feet wide, use 2 x 8s; if it's up to 6½ feet wide, use 2 x 10s; and up to 8 feet, use 2 x 12s. This seems like overly heavy construction, but the headers must support a roof or second floor or both. Extra-wide windows include picture or double windows.

Nail a 2 x 4 between the supporting studs at the height of the bottom of the window's rough opening. Short studs, called cripple studs or cripples (another mysterious building term), are nailed on 16-inch centers to support this crosspiece, which is the sill of the rough opening. Toenail the sill to the supporting studs with 10d nails; face-nail the sill to the cripples with 16d nails, and toenail the cripples to the floor plate with 10d nails.

You should do the door and window openings before the wall is erected because you can do part of the rough openings while the wall is on the floor, and part of them when the wall is up. In fact, it's a good idea to continue the floor plate along the rough openings of doors so the wall will stay in one piece while you raise it into position. For instance, you can install the full-length studs for door and window openings while the wall is on the floor. Once the wall is in place and floor plate nailed, you can install supporting studs, headers and sills, plus the cripple studs. This will help make the opening plumb and level. Before putting in the supporting studs, cut the floor plate off at the point of the full-length studs. Then the supporting studs go right to the subfloor.

Of course, where you have a window opening, the floor plate remains as a base for the cripple studs. Another thing about window

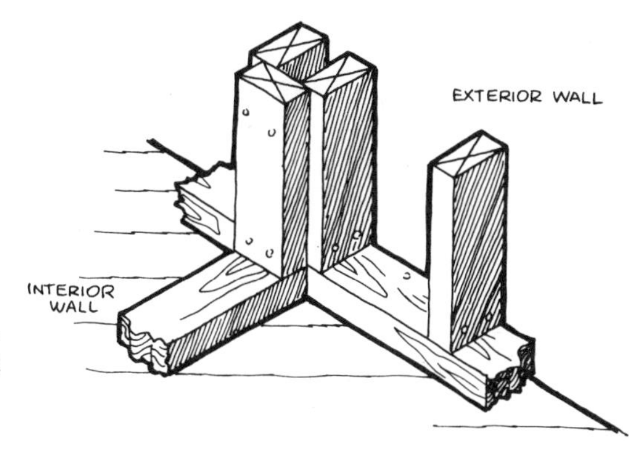

FIGURE 20. Interior treatment where two walls intersect, to allow nailing surfaces in the corners and for all walls.

openings. If you double the 2 x 4 sill (bottom of the rough opening) you'll add strength to the wall and window. And when installing cripple studs, nail one next to each supporting stud; what you're doing is making a cripple stud at each end of the sill.

If you have only a two-man crew, you can build the walls in short increments. For instance, if the length of the house is 30 feet or more, it would be a good idea to build the walls in 16-foot increments, so that you and your partner won't get a hernia putting up an overlong and heavy wall.

Also, when you build a stud wall, and dutifully put the studs on 16-inch centers, the last stud next to a window or door opening or the end of the wall might be less than 16 inches from the final stud. This can't be helped. But don't try to compromise and leave an extra-wide opening that is more than 16 inches. Neither should you try to center the next-to-last stud in the extra-wide opening. In each case you can run into trouble when applying sheet material, both inside and out.

Now the wall is ready for raising. Just put the wall on its floor plate, with the plate running along the edge of the subfloor. Nail the floor plate to the subflooring with 16d nails, making sure they go into the floor joists or header or stringer joists. One nail for each joist is enough.

Plumb the wall (make sure it's vertical) at both ends and along its side (plumbing must be done in two directions) and secure with a temporary 1 x 6 or 1 x 8 brace by nailing the brace diagonally, from near the top of a stud to a 2 x 4 cleat nailed to the subfloor. Braces can be every 10 feet or so, or where convenient.

Then build the other three exterior walls and erect them in a similar manner.

Now comes the tricky part. Nail the corners together, making sure that both walls are plumb. You may have to move the braces, which stay on until the sheathing and other framing is done. Nail corners with 16d nails. You can also nail up headers, supporting studs, sills, and cripple studs for rough openings at this time. But before all the detail work, nail the upper top plate onto the lower top plate, lapping the top plate over the lower top plate at the corners. This helps tie the walls together. Use 16d nails on 16-inch centers, two nails where the plates overlap.

Huzzah! You've come a long way.

You've got the outer shell frame up, and now must worry about interior partitions, which not only help hold up the second-floor joists but also tie in the entire wall system. Before tackling the interior walls, a system by which the interior walls attach to the exterior walls must be made, as well as a system of nailing interior wall covering.

It's not that mysterious and there are three methods to choose from. You can double the studs in the exterior wall (just 2 inches of space will do it) where the interior wall will butt against the exterior wall (Figure 20). Then nail the end stud of the interior wall with 16d nails to the doubled studs. This automatically makes a nailing surface in the corner — each corner, in fact — for plasterboard or other material. Another method is to nail 2 x 4 horizontal blocking between studs where the interior wall will interact with the exterior wall. To connect the studs, nail the blocking every 16 inches, and recess it ¾ of an inch. Then nail a 1 x 6 vertical nailer onto the blocking. The end stud of the interior wall is nailed to the 1 x 6, into the 2 x 4 blocking. Use 16d nails for the blocking, 10d nails for the 1 x 6. A third method is to nail a vertical 2 x 6 in place of the 1 x 6 and blocking.

All places where interior walls intersect and meet must be set up this way to provide a nailing surface in corners.

Also, on exterior walls, 2 x 4 blocking should be nailed with 16d nails between studs, midway between floor and ceiling. These are fire stops, and while fiber-glass insulation or any nonflammable insulation will act as a fire stop, many codes still require the blocking.

Supporting interior walls (where second-floor joists will be set, just as the first-floor joists were set over the center beam in the cellar) must be made of 2 x 4s. Nonbearing walls can be 2 x 3s, but generally 2 x 4 studs are used throughout the structure. Interior walls are built on the floor and raised into place. The upper top plate is nailed on last, overlapping the lower top plate to help hold the structure together.

So far the chronology of steps in building a house has been routine. But in some cases from now on the steps may be different, or at least in a different order, according to the experience of the crew and the number of workers involved.

With a good-sized crew on the job, framing continues on the second floor and roof while sheathing is started. Sometimes a crew will sheath a wall before it is erected.

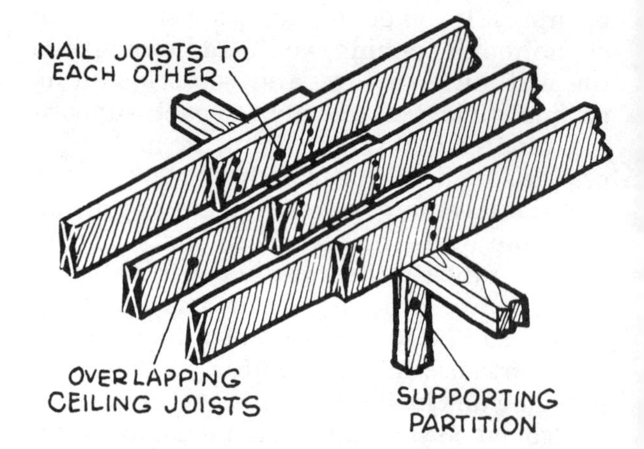

FIGURE 21. Second-floor joists (first-floor ceiling joists; the terms are interchangable) overlap over interior partition and are nailed together at each of their ends. The same technique is used over the first-floor center beam.

The simplest thing for the amateur to do, unless he has professional help, is to sheath the wall of the first floor first, then the second floor, even before the roof rafters are up. This will give extra strength to the structure, the most important consideration during construction. Sheathing is discussed in Chapter 10.

The installation of second-floor joists, subfloor, and second floor walls is easy to describe, because it is the same as everything that was done on the first floor. The edge or stringer floor joist is toenailed to the top plates of the first floor studs. The joists go in the same direction as the first-floor joists, because they cover the same span and are supported in the center by load-bearing partitions, which are at right angles to the joists both above and below.

Regular joists are strung from top plate to center partition, and are overlapped where they meet, over the center partition (Figure 21). They are toenailed to the top plate of exterior and partition walls with 10d nails. Overlapping joists are spiked together with four 16d nails.

Remember, the ends of these joists are set back 1½ inches from the edge of the top plates, to make room for the header joists, which will connect all the joist ends. In fact, the header joists can be placed first, toenailing them with 10d nails.

Again, the joists are first laid without nailing, so they can be notched or shimmed if higher or lower than their neighbors.

The subfloor is laid with proper openings (stairwell, chimney, and fireplace), and the second-floor stud walls are erected.

The only difference in a full second-floor joist system is that when you have a garrison type house, the second-floor joists overhang the front wall by one foot (Figure 22). The length of the overhang is at right angles to

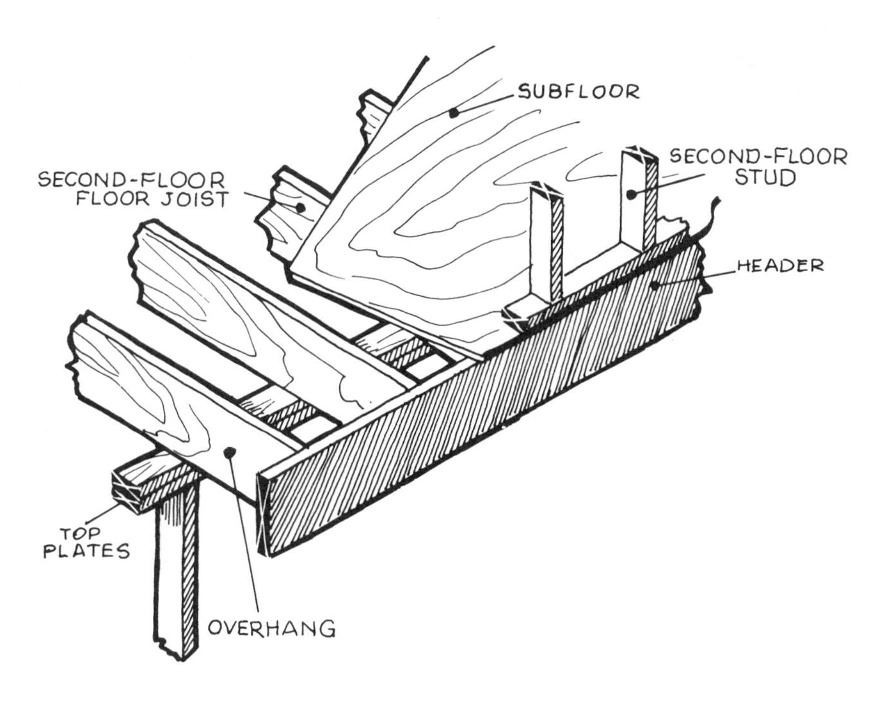

FIGURE 22. Overhang of second-floor joists, as in a garrison-style house. The overhang is a cantilever.

the joists, so the joists are simply made longer and allowed to hang over the top plate by 10½ inches, and connected with a header joist.

Where the overhang happens to be where the joists are parallel to the length of the overhang, the parallel joists stop about 6 feet from the outside of the top plate at the overhang. This last joist is doubled. Then joists 6 feet 10½ inches long are attached with joist hangers to the doubled joist and allowed to overhang by 10½ inches. The 1-foot overhang is made possible by nailing a header joist on the ends of the lookout joists.

One more word about those interior partitions. You not only need nailing surfaces for wall covering (plasterboard or lath for plaster), you also need horizontal nailing surfaces where ceiling joists (second-floor joists; the term is interchangeable) meet the walls.

The technique is similar to the vertical nailing surface systems. Where the partition wall parallels the joists, you can put 2 joists about 2 inches apart, with half of each bearing on the 2 x 4 top plate. This leaves ¾ inch on each side of the partition for nailing.

Where joists are at right angles to the partition wall, install 2 x 6 blocks between the joists, nailed directly on the top plate and toenailed to the joists. This will leave 1 inch on each side of the partition for nailing. These techniques are essential where joists run parallel to the partition wall. Otherwise, you'd get ceiling material ending in midair where it meets a wall.

Where the ceiling joists meet the exterior wall, the techniques are the same, except you need only to provide a nailing surface on the inner side of the wall. Where joists are at right angles to the partition wall, some people don't bother with a nailer, allowing ceiling material to span the joists at 16-inch centers. This is not too bad a trick, except the other way is found in quality construction. And the other way gives a good, solid corner.

Some building codes require corner bracing (Figure 18), which is simply a 1 x 4 or 1 x 6 attached to the studs diagonally from the top of the stud at the corner and slanting to the floor plate at a 45-degree angle. It must be let in, which means that the studs and plates must be notched to receive the board and make it flush with the stud surface. Nail with two 8d nails in each stud. The corner brace may not be needed if certain sheathing materials are used and techniques are practiced (which are explained in Chapter 10).

On with the sheathing!

chapter 10

Putting the Flesh on the Bones

Exterior Sheathing

Sheathing is any covering that goes on the outside of the wall studs, floor joists, and roof rafters. If a subfloor wasn't called a subfloor it would be called floor sheathing. Basically, the sheathing is the fleshing out of the skeletons — the bones, if you will — of the framework (Figure 23).

The amateur should sheath the outside of the first-floor framing before he goes any higher with floor and wall framing. The sheathing will make the structure considerably more rigid, and will help tie the wall into the floor framing, on both first and second floors.

There are several different kinds of sheathing. Wood sheathing (boards), plywood, and structural (fiberboard) insulation board. Over all these materials goes siding (shingles, clapboard, etc.), which makes the house weatherproof. Another technique, if codes allow it, is to install a combination sheathing and siding board. This is usually a ⅝-inch-thick plywood sheet 4 x 8 feet and grooved to simulate vertical siding. Sheathing boards, plywood, or structural insulating board are recommended for sheathing.

Boards are nominally 1 x 6, 1 x 8, or 1 x 10, and can be matched (tongued and grooved), shiplapped, or square edged. Spruce is usually used, but other species, such as fir, southern pine, and hemlock, are sometimes available.

Plywood sheets, usually 4 x 8 feet, come in thicknesses of 5⁄16, ⅜, ½, and ⅝ inch. For best results, particularly if you use siding such as shingles, where the nails must hold in the sheathing and not in the studs, use ½-inch exterior-grade plywood. Some plywood comes in lengths greater than 8 feet, but it might not be available and is tougher to install.

Insulating board is made of fibers and comes in three types: regular density, intermediate density, and nail base (able to hold a nail). They are treated to be water-resistant. Thicknesses are ½ inch and 25⁄32 inch. The nail base insulating board is your best bet. And if you use insulating boards, it is still a good idea to use regular insulating materials in the wall cavity. There is no such thing as too much insulation, although there is a practical limit, and there is a point where additional insulation is not worth the expense. However, with insulating board and regular cavity insulation, you have not hit that point. Insulation is as important in warm climates as in cold, to reduce costs of air conditioning.

The thickness of the sheathing is impor-

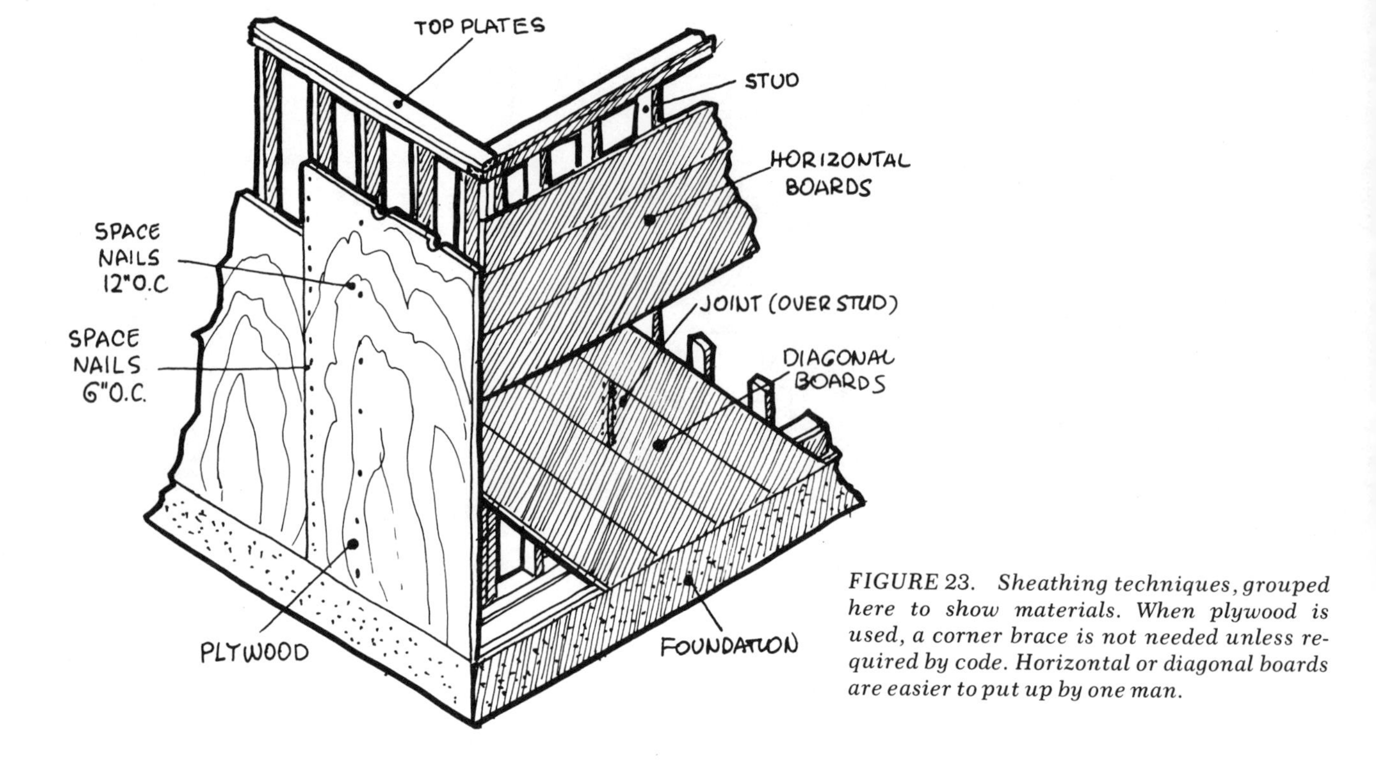

FIGURE 23. Sheathing techniques, grouped here to show materials. When plywood is used, a corner brace is not needed unless required by code. Horizontal or diagonal boards are easier to put up by one man.

tant because of the standard thickness (depths) of the jambs in windows and doors. For instance, with a 2 x 4 stud (3½ inches), ½ inch sheathing, and ½ inch plasterboard on the interior, you get 4½ inches, and must specify this size for the jambs.

Boards can be laid up horizontally or diagonally. Use 8d nails, two in each stud crossing with 1 x 6 boards and three for wider boards. Three nails must be used for all diagonal boards.

One man can handle boards more easily than plywood, which is not only heavy but awkward to handle. Matched boards are best to use. When a board is slightly curved, and nailing doesn't close up the gap while it is being installed, here's a trick: drive two nails at a downward angle through the tongue. The nails add power, usually enough to snug the board down on its neighbor. Then finish with face-nailing.

Horizontal sheathing is the easiest to apply, and less wasteful than diagonal application. Joints should meet at a stud, and no two joints should be contiguous. Corner braces must be used with horizontal boards.

Diagonal boards are installed at a 45-degree angle, and corner braces are not necessary unless required by code. Diagonals can be started at a corner and ended at half the length of the wall, with the diagonal going in the opposite direction for the second half of the wall.

Now, here is where sheathing helps tie in the entire structure. In Chapter 8 we described how the sill is set ¾ inch from the outer edge of the foundation, to make room for sheathing. When diagonal sheathing is put on the studs, it will not only span the studs but reach down to the sill as well, including the perimeter floor joists (stringer and header joists), tying the sill, floor, and wall into one unit (Figure 23).

You can tie in the second-floor joists and wall easily with plywood sheathing. Half-inch sheathing is recommended. If you offset the sill by ¾ inch on the foundation, and apply ½-inch plywood, don't worry about the ¼-inch shelf remaining on the foundation. It will be covered adequately by the siding.

Plywood sheets should be applied vertically, using 8d nails spaced 6 inches at edges and 12 inches at intermediate studs. Some people feel safer nailing every 4 inches at edges and 6 inches at intermediate studs.

With vertical plywood, set the 4 x 8 foot sheets on the lip of the foundation, spacing the sheets 1⁄16 to 1⁄8 inch to allow for expansion and contraction due to moisture content. Because the height of the wall is 96 inches, plus 3⁄4 inch for the subfloor, 9½ for the floor joists and 3½ for the sill, you get a total of 109¾ inches. All this fiddling around with figures is simply to show that a 96-inch sheet of plywood will be 13¾ inches short of the top of the top plate of the wall. This is good, because when you put on the second-floor joists, floor, and wall, the next succeeding plywood sheet will tie in the first-floor and second-floor walls perfectly. This tie-in is particularly important in areas of high winds. Otherwise special metal ties must be installed to tie the wall into the sill.

Structural insulation board is, in our opinion, not as satisfactory as plywood or boards. But it can be used, according to building code requirements, nailed with galvanized roofing nails, spaced 3 inches at edges and 6 inches on intermediate studs. When structural insulating board is used, and the material is nail base, special ring-shanked nails must be used for nailing siding into this material. Clapboards can be nailed at every stud. When the insulation board is not nail base, clapboards must be nailed at every stud. Shingles must have 1 x 3 nailing strips nailed to the studs. They must be spaced horizontally to correspond with the spacing (exposure to the weather) of the shingles. Sheathing should be covered with a membrane that is water-resistant but not necessarily vapor-resistant. Use 15-pound roofing felt (the familiar tarpaper) or slater's felt, a lighter and cheaper tarpaper.

There is some controversy over using sheathing paper that is not vapor-resistant. One side argues that the sheathing paper should be vapor-resistant, and recommends aluminum foil. The other side argues the other way, maintaining that wood sheathing should be allowed to breathe, so recommends a paper that is not vapor-resistant. We stick with the "breathing" theory, and maintain that, generally, aluminum foil has little or no reflective value unless there is an air space between it and the siding.

Whatever kind of paper you use, it should go over sheathing and under siding, including stucco (a concrete plaster applied to metal lath) and brick veneer. It does not have to be used over plywood or other sheet materials considered to be water-resistant. The sheathing paper is applied horizontally, with 4-inch laps.

The point of sheathing paper is to add to the water-resistance of the wall and, in the case of windows and doors, to make the joint between the casing and sheathing resistant to the passage of air, which is accomplished by using 8-inch strips over and beside window and door openings (see Chapter 15).

chapter 11

Time to Raise the Roof

Roof Framing

Now you have two full stories framed and sheathed, including a first-floor platform on the foundation, plus first-floor walls, plus second-floor platform and second-floor walls.

It's time to install ceiling joists and roof rafters (Figure 24). Ceiling joists, no different from floor joists except in size, support a ceiling rather than floor.

Roof rafters are set at an angle, reaching from the top plate to ridgepole or board, and are called rafters because they are at a relatively steep angle. For a gently sloping roof, particularly if there is no peak or ridge, the rafters are called roof joists.

First on the schedule are ceiling joists which, if they are supporting simply a ceiling and not a floor, as in the case of the garrison Colonial we are discussing, can be 2 x 8s instead of the 2 x 10s that were used for the first and second floor. They are laid up in the same way as floor joists; that is, two joists spanning the distance from the top plate of the second-floor wall to the opposite wall, overlapping in the center, over a bearing partition wall. The nailing is the same as for the floor joists.

Several details are different from floor joist construction, however. There are no header joists connecting the ends of the joists. Nor is there a stringer joist at the outside of the top plate. And the ceiling joists must be cut at an angle at their ends to accommodate the rafters. These details will be described later, but now the problem arises: what angle should the rafters be set at, to make what kind of a pitch?

Roofs are described as pitched or flat. The pitched roof usually has a relatively steep angle, and is usually peaked, that is, has a ridge from which the roof slopes down on both sides (Figure 24) A flat roof is not really flat, but rather is low-pitched and slants in one direction, like a shed. A truly flat roof is not practical for the amateur. Because we are talking about a garrison Colonial house, we will confine ourselves to a simple pitched roof.

Pitch is the slant of a roof, determined by the rise, which is the vertical distance from the top of the wall top plate to the top of the ridgeboard. The span is the entire width of the house, and the run is half the span, from top plate to the midpoint, directly below the ridgeboard.

So, what angle to pitch the roof? This will be detailed in the plans, but the length and cuts in the rafters are determined by the pitch.

FIGURE 24. Roof construction, showing end studs that are notched to accommodate end rafters. Detail shows how ceiling joist is set back from rafter to allow seating of end studs. It is secured with a spacing block.

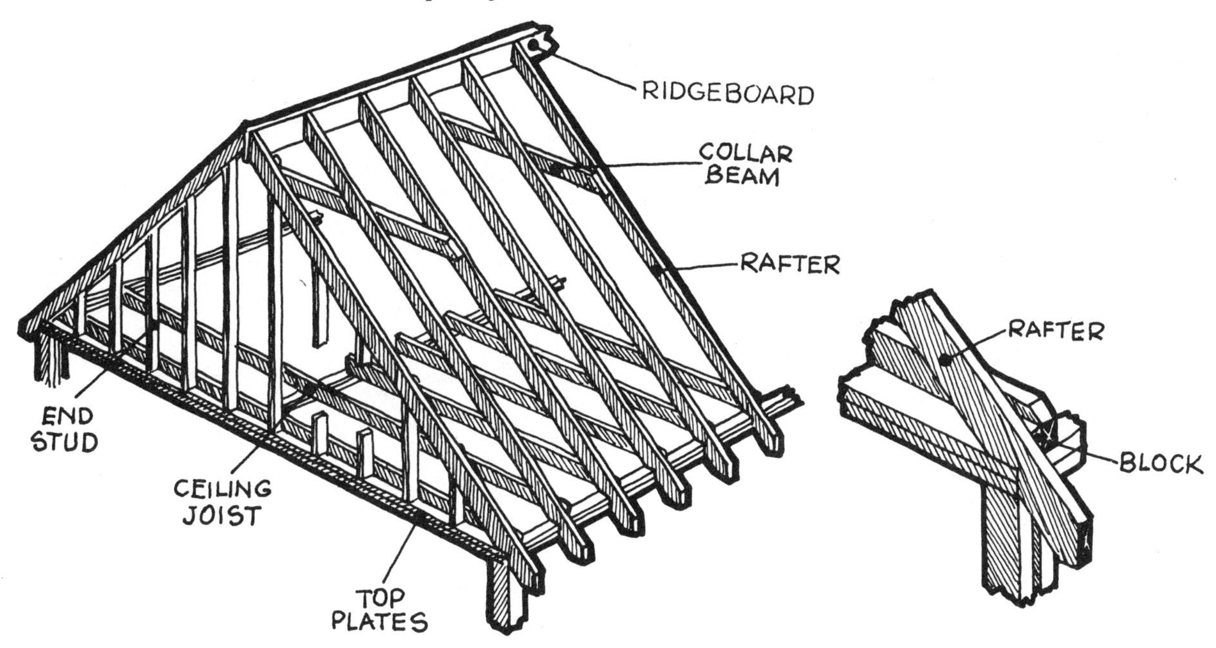

The pitch depends on several things: design, ability to throw off snow, and ability to hold or resist the sun's heat. A steep roof will throw off snow, and will absorb more of the sun's rays, which are at a low angle in the winter. With today's maximum insulation, both against heat and cold, our main concern is with ability to throw off snow in heavy snow areas.

The pitch of a roof is determined by how high it rises to the ridgeboard. An ideal slope is 45 degrees, so every cut is made at a 45-degree angle. But this may make the roof too steep or too high for a two-story house. It is ideal for a Cape Cod house roof, which is built up from the top of the first-floor walls.

Say you want a 45-degree pitch. The house is 30 feet wide, which means the span is 30 feet and the run is 15 feet. This in turn means the rise should also be 15 feet, because a 45-degree pitch rises 12 inches for each 12 inches of run. A 45-degree pitch is also called a half pitch. This is too high and too steep, so let's make it a 10-foot rise to a 15-foot run, which is an 8-inch rise to each 12 inches of run. This is called a one-third pitch because the angle of the pitch is 30 degrees, or one-third of vertical.

Now, how do you determine the length of the rafters, which will be longer than the run, and the angle to cut them so they will butt against the ridgeboard?

This is where the framing square comes in, one of the most useful tools you can buy. It is a large L-shaped square, with one arm 24 inches long called the body and a shorter arm 16 inches long called the tongue. Each arm is marked in inches and is calibrated to indicate the length of the rafter and angles to be cut.

Length of rafter can be determined by a simple algebraic formula: $A^2 + B^2 = C^2$. "A" represents the run (15 feet), "B" indicates the rise (10 feet), "C" is the rafter, which is the hypotenuse of a right triangle, which is always longer than either of the other two sides. Put another way, the rafter length squared is equal to the square of the run plus the square of the rise.

So, square the run ($15 \times 15 = 225$), add it to the square of the rise ($10 \times 10 = 100$) and you get 325. Take the square root of 325 and you get 18 ($18 \times 18 = 324$, close enough). So the rafter length is 18 feet.

This type of formula has already been worked out, and this is where the framing

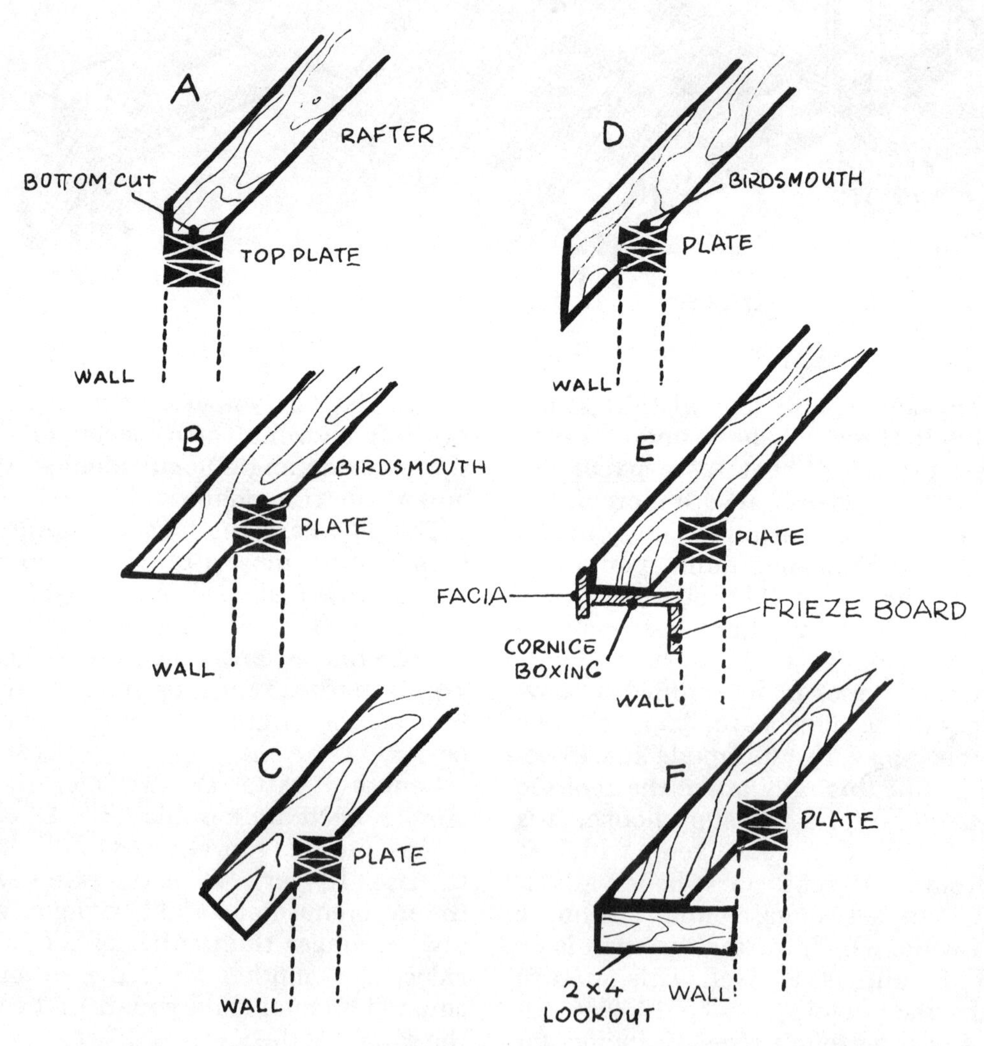

FIGURE 25. Rafters. A, close cut; B, level cut; C, square tail (no cut at all); D, plumb cut; E, plumb and level cut covered with facia and soffit to form cornice (boxed eave); F, wide eave overhang using 2 x 4 lookouts to form cornice, which is then covered by facia and soffit.

square comes in. If you place the square on the rafter, body horizontal and tongue hanging down, with the rafter at the approximate angle of its intended position, you can line up the rise (10) on the tongue and the run (15) on the body and measure the distance between the two points. It will be 18.

You can make this measurement on the square even without a rafter. If you match the numbers 10 and 15 (outside or inside), the length from mark to mark will still be 18. So once you get the hang of it, it's foolproof. But it has to be tried personally. No amount of talk or writing will make it clear until you try it.

So the rafters should be 18 feet long, minus ⅜ of an inch to compensate for half the ridgeboard, which is ¾ of an inch thick. If you use 2 x 8 rafters, use a 1 x 10 ridgeboard. The extra width of the ridgeboard will correspond to the extra width of the rafter cut at an angle.

But don't cut the rafter yet. Add to its length enough so you'll have an eave overhang, which is determined by the design and style of the house. A traditional or Colonial house has a short overhang (6 to 10 inches or so). A modern house usually has a wide overhang, up to 3 feet.

The eave end of the rafter can be cut plumb, (Figure 25A, D), which means its end will be vertical when the rafter is in position. It also can be cut level, (Figure 25A, B), which means the bottom of the eave will be at right angles to the plumb cut, or horizontal. If neither cut is made, you'll get a square tail on the rafter (Figure 25C), which is easier to do but is not good design.

Now, measure the rafter so it will be 17 feet 11⅝ inches long, and add for the eave overhang, which will be discussed in Chapter 13. Now make the angle cuts, by placing the framing square with the run and rise numbers in the right position, marking it with a pencil, and cutting it. This is done at both ends.

You're still not ready to put the rafters in place because if you connected them on the ridgeboard and let them set on the top plate, the contact on the top plate will be inadequate, and nailing would also be inadequate.

So you make a notch, or bird's mouth (Figure 25), using the same angle for the vertical cut as you did at each end of the rafter. The horizontal cut must be at right angles to this vertical cut. Now the rafter will sit on the top plate and can be toenailed to it and face-nailed to the ceiling joist beside it.

Now we get back to the ceiling joists. Their ends are cut at the same angle as the pitch of the roof rafters so they will not protrude above the tops of the rafters (Figure 24). The ceiling joists go up, as we said, the same way as the floor joists. However, the joists at each end of the house, which are the gable ends, are set 3½ inches back from the outer edge of the top plate. Yessir, the two end joists are set just at the inner edge of the top plate (Figure 24). This is to allow gable-end studs, for the wall of the gable end, to sit on the top plate.

First, install the ceiling joists. Toenail them to the top plates with three 8d nails, two on one side and one on the other. Same goes for where they hang over the top plate of the interior bearing partition. Where the joists butt, nail with four 16d nails.

Amateurs are likely to have nightmares

over how to get the ridgeboard and rafters up. It is pretty tough without an experienced crew. But there is one way to do it without a large crew and without a lot of experience. Place the end rafters right at the end of the top plate, so their face is even with the face of the wall studs. Place a 2 x 4 block between the end rafters and end joists. But instead of putting up the end rafters first, put up the gable end studs first (Figure 24).

These set on top of the top plate, on 16-inch centers (their length depends on the position of the end rafter), and are notched at their upper ends so they can be nailed to the rafters. They must be cut to graduated heights to fit the pitch of the rafter. You can cut the highest pair of studs to the right height and toenail them to the top plate. Brace them for support.

Then install the end rafters, leaving a ¾-inch gap for the ridgeboard. Then, with the rafter in place, you can measure the other gable end studs, notch them, and nail in place. Use three 8d nails for toenailing and three 8d nails for connecting stud to rafter.

Once you've installed the gable-end studs and the pair of facing end rafters, with a gap at the peak to accommodate the ridgeboard, you can sheath the gable wall with whatever sheathing you choose. Once the gable end is sheathed and braced so that it is plumb, do the same at the other end. Now, you can cut the ridgeboard to the exact length of the house and connect it to the end rafters. If you can't get a ridgeboard long enough, build it in two pieces, with their ends butting where two rafters meet. The point of doing the gable-end walls first is to give a good base for applying the rest of the rafters.

Because you'll be working on the ceiling joists, nail a temporary catwalk of 4 x 8 sheets of plywood as a floor. You can nail the rafters to the ridgeboard from this floor. The low ends of the rafters are best worked on from a ladder.

Once one set of rafters is cut and fits right, you can use them as templates for the rest of the rafters.

Mark 16-inch centers on the ridgeboard and nail each pair of rafters onto the ridgeboard and top plate. Face-nail through the ridgeboard and into the rafter ends, using three 10d nails. Toenail rafter to top plate with two 8d or 10d nails. Face-nail rafter to ceiling joist with four or five 12d nails.

To keep the ridgeboard straight, nail each pair of rafters opposite each other, rather than all rafters on one side. When nailing the ridgeboard to the second, opposite rafter, you'll have to nail at an angle, because the first rafter, already in place, will be in the way. This is standard procedure.

Openings for chimneys, fans, and skylights must be built as the rafters are installed, using the same technique as openings in floors; that is, doubling the rafters at each edge of the opening and doubling the headers connecting the rafters at the opening (Figure 26). Joists must be 2 inches away from any masonry; sheathing ¾ inch.

To keep rafters from spreading (there is great pressure at their ends), install collar beams (or collar ties) connecting the rafters no farther up the rise than halfway (Figure 24). The rafters form an upside-down V; the

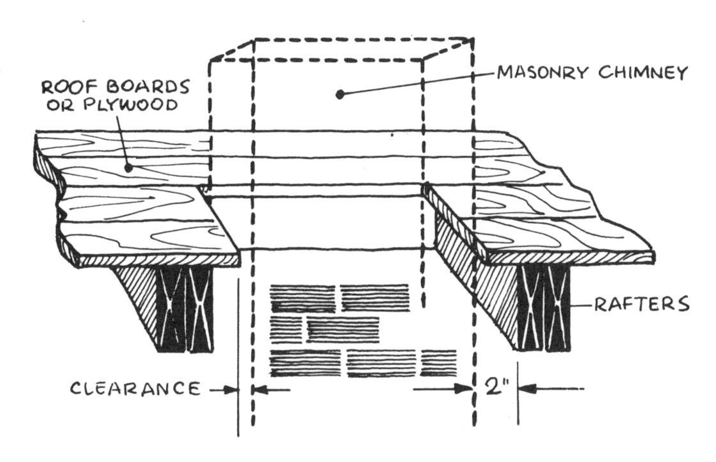

FIGURE 26. Opening for chimney has doubled rafters. Two-inch clearance is required between masonry and framing pieces (rafters); ¾-inch clearance between masonry and sheathing.

tie makes each pair an A. If the attic area is to be used as living space, as in a Cape Cod house, the collar ties can go as high as 7 or 7½ feet and will also be ceiling joists.

Make the collar ties the same size as the rafters (2 x 8) and if they are also ceiling joists, nail one on every pair of rafters, using four 12d nails. In an attic where living space is not planned, the ties can be secured to every fourth pair of rafters.

Overhang at the eaves acts as a drip edge, and protects the siding. The use of gutters has reduced the need for such overhangs, but they (the overhangs) are a part of good design (Figure 25 E, F). A small overhang or none at all makes a house look skimpy (Figure 25 A).

Many modern houses, and those in warm climates, have very wide overhangs, up to three feet, to keep the sun off the walls for a cooling effect. Even in northern, colder climates, they serve the same purpose in summer. And in winter, when the sun is at a lower angle, the wide overhang does not prevent the sun from warming the walls and windows.

We mention eave overhang here because the overhang is determined by the length of the rafters.

As for the overhang at the rake (gable end), that's a different matter. Traditional and Colonial houses have very little overhang at the rake. Usually it is just a nominal 2-inch facia (see page 82) block nailed under the roof sheathing and into the wall sheathing, and covered with a facia or rake board. Roof sheathing is extended over the facia block and rake board. A narrow overhang is called a close rake. Close rake and narrow and moderate rake overhangs (up to 12 inches) are discussed in Chapter 13.

For rake overhangs 12 inches and wider, modification of rafters is required. The end rafter is eliminated, and the next rafter is doubled. Gable-end studs are cut at the same angle as the rafters and installed, and covered with a 2 x 4 plate, so they are 3½ inches (the width of a 2 x 4) below the high side of the rafters.

Then, 2 x 4 lookouts are extended from the doubled rafter over the 2 x 4 plate and allowed to cantilever (look out) beyond the gable wall as planned. A fly rafter is nailed to the ends of the lookouts, connecting them. The lookouts rest on the plate directly above the gable-end studs and are attached to the doubled rafter by metal joist hangers. Nailing blocks of 2 x 4s or 2 x 6s are nailed on the plate, overhanging it by 1½ inches to act as a nailing surface for the soffit (undertrim for the overhang). Trim detail is described in Chapter 13.

Overhang for low-pitched roofs is quite a bit easier than on high-pitched roofs. Incidentally, the wider the overhang, particularly at gable ends, the more expensive the construction. For low-pitched roofs, the roof joists are extended in the direction they run and cantilevered over the exterior-wall top plate as far as planned. When the joists are parallel to the wall, lookout joists are extended from an interior doubled joist to make the overhang. Three feet is the maximum overhang possible without getting into specialized construction.

For some modern construction, roof rafters are prebuilt and called trusses or trussed rafters. Each truss is built of nominal 2-inch lumber and held together with steel or plywood gussets. They are designed to span an area of 20 to 32 feet, and allow large areas to be free of supporting members except from outside wall to outside wall. Then, interior partitions can be placed without regard to supporting the ceiling joists and rafters. Roof trusses need a larger crew and usually a crane to lift into place. They also usually require additional tiedowns such as metal anchors and cleats, in addition to regular toenailing.

A garrison Colonial roof is not designed to have dormers because the attic is not occupied. However, on a Cape Cod type house, where the roof begins at the top of the first-floor wall, dormers are an integral part of the construction, providing light, cross-ventilation, and more room in the second-floor living space.

A common dormer (from Old French *dormeer,* "bedroom window," from *dormir,* "to sleep") on Cape Cod style houses is the shed dormer, named after its shed roof, which has a shallow pitch in one direction (Figure 5).

Although this book describes in general the building of a garrison Colonial, straight two-story gable-roofed houses can be dull if there aren't variations, so here are some of those variations, at least in the roofs.

A shed dormer adds a lot of space to a second floor covered by a double-pitched roof, both in headroom and in floor space, by allowing headroom all the way to the edge of the house.

Shed dormers are best built on the back of a house, because they are not particularly handsome, and give a Cape Cod style house a top-heavy look if they're in the front. And if the house is situated so its gable end is facing toward the street, a shed dormer can give a lopsided look. But its advantages can outweigh these style objections.

Let's talk about a full shed dormer. Instead of rafters on the dormer side of the roof pitched at the same angle as those on the opposite side, the rafters slope at a very shallow angle from the ridgeboard, and attach to a full wall set on the second-floor platform. The wall is built in the same manner as any stud wall, with a floor plate and a doubled top plate. The rafters extend over this wall at a shallow angle, and the length of the extension is determined by the width of the overhang. Rafters are measured and cut at the proper angle in order to sit

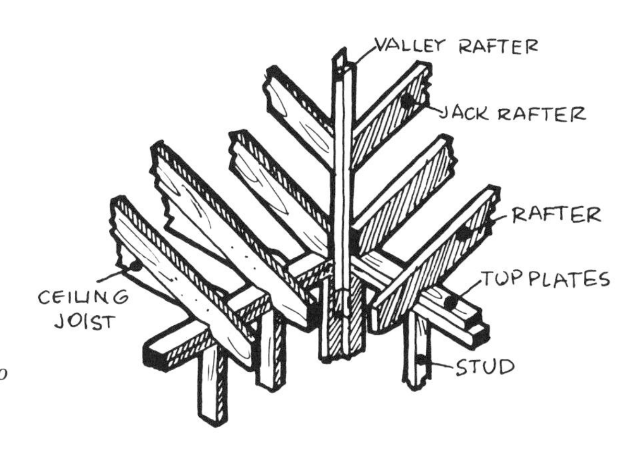

FIGURE 27. The valley rafter is doubled when two sloping roofs intersect to form an "inside" corner.

against the ridgeboard, and are notched in order to sit onto the top plate. End studs are installed in the same way as the gable-end studs (see page 70).

It is good practice to install a rake board along the dormer side of the gable-end wall to correspond with the rake board of the roof opposite the dormer. This brings the gable end to its proper perspective and avoids a huge expanse of uninterrupted siding.

Sometimes the dormer sides are set two or three rafters in from the end rafters, to give better proportion to the dormer. The resulting space where the dormer is not in place is not wasted, though. The space allows plenty of room for a closet, which for proper clothes hanging does not have to be more than 2 feet deep. If the dormer starts two rafters in, the width of the closet is about 32 inches. If the dormer starts three rafters in from the edge, the rough closet is about 48 inches wide.

Other dormers, called eye dormers (Figure 5), are installed in the front of a Cape Cod style house for light and ventilation. They can have a shed or gabled roof, and dormers with the latter type of roof are called "A" dormers. They also give the front of a Cape Cod house more style, relieving the big expanse of a slanting roof.

A gabled roof dormer is installed the same way as a dormer built two or more rafters in from the end rafters. The side walls are installed over doubled rafters. However, they usually don't extend to the edge of the house proper, and if you looked out of an eye dormer's window, you'd see below you a bit of roof slanting at the same pitch as the main roof. Where these short rafters stop at the wall of the dormer, the header is doubled.

And where the gabled roof hits the slant of the main roof, the construction gets pretty tricky, requiring short rafters to be set at an angle to the other rafters. Where the two roofs join, you get a valley on each side of the dormer roof. The rafter that makes up this valley is called, appropriately enough, a valley rafter. The short rafters are called jack rafters (Figure 27). Their ends must be cut at the proper angle in order to sit against other rafters.

Which brings us to the many variations that a roof can have. On a hip roof the roof line comes down to the eaves on all four sides of the house, eliminating the gables. It is usually employed where the two full stories of a house are particularly tall or large, and eliminates that "soaring" look. A hip sounds terribly complicated, and well, it is, a little. The angle required for jack rafters (Figure 28) is 45 degrees. A hip roof can have a horizontal ridge, long on a rectangular house, short on a square one. The rafters butt against this ridge in the regular manner. Then, a hip rafter extends from each corner of the house to one end of the ridge. The angle of the hip should be the same angle as that of the full-length rafters, called common rafters. If it isn't, the end cut must be figured out with the rafter square so it will butt properly against the ridgeboard.

Then, short jack rafters, called hip jacks, are cut to extend from the top plate of the wall to the hip rafters. The hip rafter must be 2 inches deeper than the common rafters in order for the jacks to pay fully against the hip rafter, because jack ends are cut at an angle, making their contact point greater than their width. Nail jacks to rafters with three 10d nails, and toenail them to the top plate with three 10d nails.

Valley rafters are necessary where two gabled roofs intersect, as in an L-shaped house. The technique of building them is the same as with hips, except in reverse.

For instance, the valley rafter (doubled and 2 inches deeper than the short valley jacks, for the same reason as mentioned for hip rafters) extends from the inside corner of the L to the ridgeboard. Jacks are nailed with three 10d nails at each end (face-nail to ridgeboard and toenail to the doubled valley rafter).

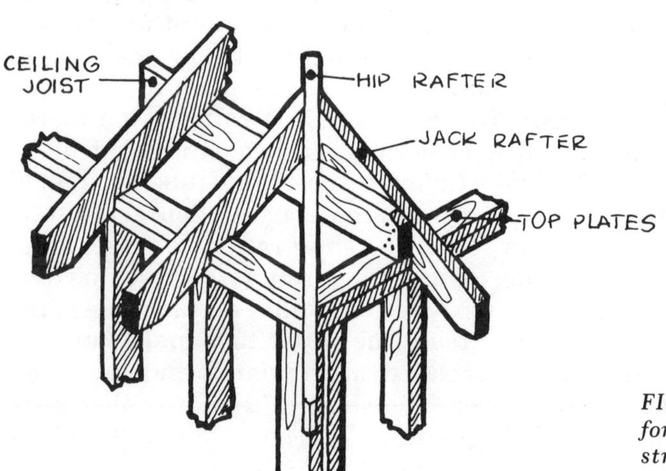

FIGURE 28. Hip rafter where two sloping roofs form an "outside" corner. Both hip and valley construction require compound cuts of rafters and can be very tricky.

The advantages of a hip roof are stability and less tendency to rack (to move from their vertical position). A hip roof also can be a pyramid roof, with hip rafters meeting at a center peak, and without a ridgeboard. A pyramid roof has no common rafters, and requires a tremendous number of double-angled cuts.

A gambrel roof, on so-called Dutch-Colonial houses, has two slopes on each side of the ridgeboard, a very steep one from the second floor to 60 to 70 percent of the total rise, and a shallow slope from there to the ridgeboard. Construction of the gambrel roof is similar to that of the regular gable roof, except the steep roof and less basic span allows the use of narrower rafters. The roof also leaves more room in the second floor, and only eye dormers, if any, are needed for light and ventilation.

The steep rafters are butted onto the top plate below, and their tops are connected with collar beams, which also act as ceiling joists. Then the top rafters are set on a ridgeboard and overlapped and nailed onto the steep rafters. The steep rafters also can end in an overhang, attached to second-story floor joists overhanging the top plate of the first-floor wall.

All except the gable roof are specialized types, and the amateur should borrow or hire the services of a professional framer if he plans a specialized roof.

chapter 12

The Final Coverup

Roof Sheathing

Roof sheathing is generally the same as wall sheathing: ¾-inch matched boards or ½- or ⅝-inch plywood or fiberboard (Figure 29).

In Chapter 13, some tricky roof-line details are described, such as the eave overhang and cornice (the boxed area at the eaves) and the rake overhang (at the gable ends). Roof sheathing is done before these details are completed, but the dimension of the details must be worked out so the roof sheathing will extend to the right point. This isn't difficult with the eave overhang, which is determined by the overhang of the roof rafters. But in the case of the rake overhang at the gables, the extent of the overhang must be determined earlier to figure how much the roof sheathing will extend over the end rafter, which has been covered with wall sheathing.

If a small crew is working, or the house builder is working alone, roof boards are easiest to apply. To keep shrinkage to a minimum, boards should be kept to nominal 1 x 6s or 1 x 8s.

Nail the boards horizontally on each rafter with two 8d galvanized nails. Threaded or ring-shanked nails also can be used, and hold very well, but the galvies are good grabbers if they are hot-dipped zinc.

The same rules apply to roof sheathing as to wall sheathing. End boards, if not matched, must meet over a rafter. And no two boards must meet over the same rafter in succession. This problem is merely one of different lengths of boards.

When boards butt each other along their long edges, the sheathing is called closed. When wood shingles are used as the roofing material (to keep the weather out), particularly in wet climates, spaced sheathing provides ventilation (Figure 29). This means boards are spaced to correspond to the exposure of the wood shingles.

For instance: if 1 x 4 nailing strips are used, and the shingle exposure is 5 inches, the space between strips is 1½ inches. The 3½ inches of the 1 x 4 strip plus 1½ inches equals 5 inches. This spacing is important so the shingles can be nailed into the strips. If the strips were not equally spaced, you might find yourself varying the shingle exposure to compensate.

Plywood sheathing obviously cannot be used for spaced sheathing, but can be used for the closed sheathing necessary with the

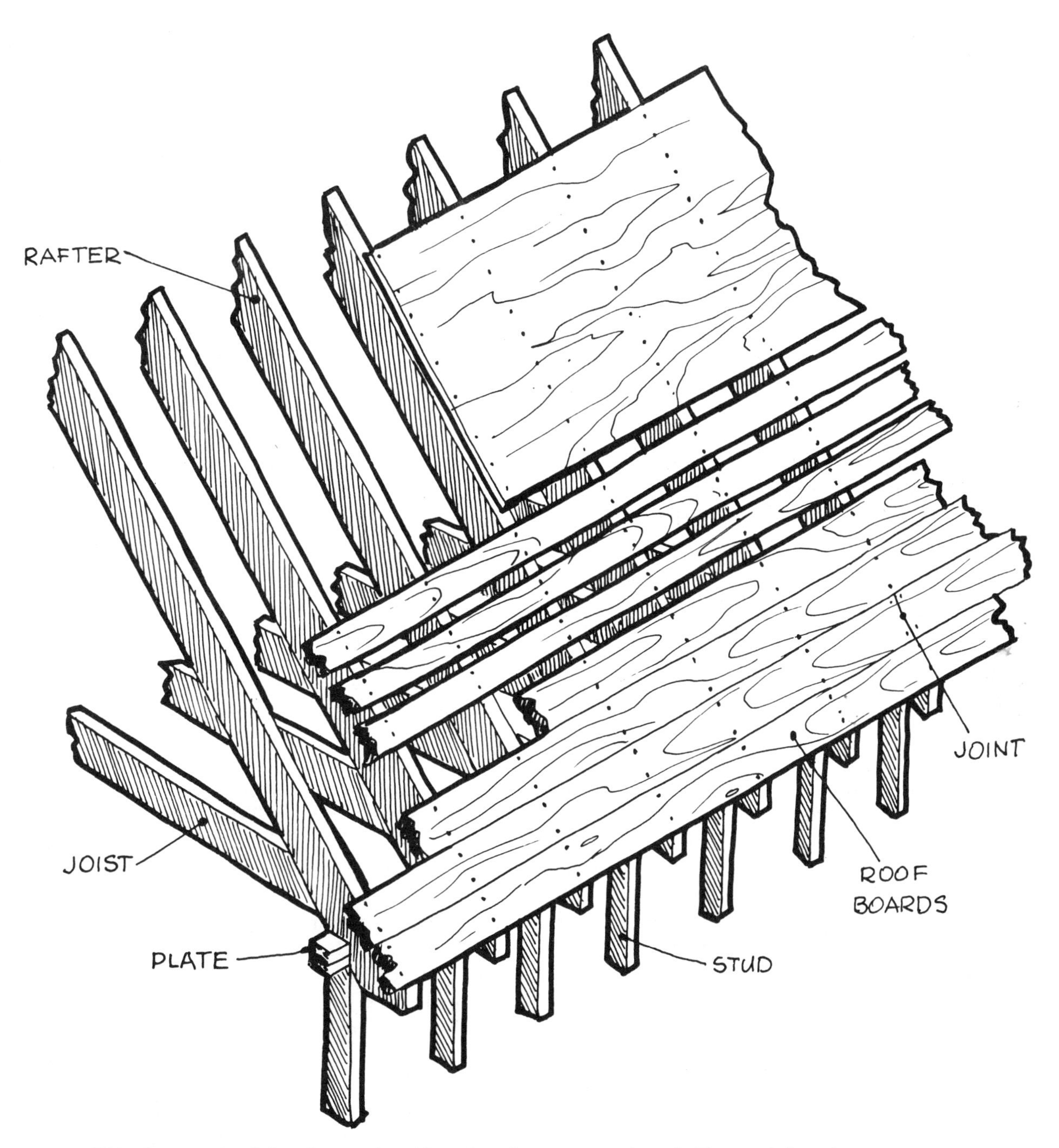

FIGURE 29. Roof sheathing. Closed boards at bottom, open boards (for wood shingles) in center, and plywood at top.

most common (and cheapest) roofing materials, asphalt shingles. Plywood can be a minimum of 5⁄16 inch, hardly more than 1⁄4 inch, which is inadequate, in our opinion, for good, strong construction and ability to hold roofing nails. Plywood 3⁄8 inch thick is better, 1⁄2 inch is best, even on 16-inch centered rafters. In fact, 1⁄2 inch is good on 24-inch centered rafters. Our recommendation is 1⁄2-inch plyscore, an exterior-grade plywood that is strong and economical (Figure 29). Even better, and necessary when slate or other heavy roofing material is used, is 5⁄8-inch plywood.

Plywood must be laid with its grain perpendicular to the rafters so it won't sag. Space joints 1⁄16 inch. Stagger butt joints so they won't meet on the same rafter. Nail plywood with 8d galvanized or threaded nails 6 inches apart on edges and 12 inches on intermediate rafters. If you feel better about it, nail 6 inches on center at intermediate rafters, too. Nails, while far more expensive than in the past, are still a pretty inexpensive item for what they do — hold the house together.

While we are suggesting traditional methods of construction, post and beam construction is growing in popularity, and even with traditional techniques a roof of deck planks (nominal 2 inches and thicker) is a natural for designs where a so-called cathedral ceiling (a corny, commercial word for peaked or slanting ceilings) or beamed ceilings are desired.

Plank roof decking comes tongued and grooved and is designed to span rafters up to several feet on center. It can also be laid from the wall top plate to ridgeboard, which is a super big beam (maybe as big as 8 by 16 inches), and act as rafter and sheathing at the same time. It is expensive to install. An advantage to this type of decking is that it acts as both sheathing and interior ceiling. When this is done, a rigid insulation is installed on top of it, then roofing shingles, using extra-long nails so they will penetrate shingles, insulation, and decking.

Fiberboard is sometimes available for roof sheathing. If it isn't readily available, it probably wouldn't pay to order it and depend on the vagaries of the American delivery system, especially if you are working on a tight schedule of material deliveries.

An important consideration for roof sheathing is the rake overhang at the gables. Sheathing boards can be extended up to 16 inches over the gable wall without sagging excessively and until support is installed. Half-inch plywood can also be extended.

For chimney openings in the roof, rafters and headers must be spaced so they will clear masonry by 2 inches. The sheathing should clear masonry by 3⁄4 inch, as fire protection (Figure 26).

If the chimney is on an outer wall, all that must be considered is a notch in the rake overhang or eave, according to the location of the chimney.

Other openings, as for skylights and ventilators, can be left rough to accommodate the ready-made units.

At ridges, hips, and valleys, the sheathing should be tightly butted and nailed. At the ridge and hip, don't worry if the sheathing doesn't butt along its entire thickness; any gap will be covered by asphalt shingles, alternated butted wood shingles, metal ridge cap, or board ridge.

At valleys, the sheathing will butt tightly at its top edge, which is proper underpinning for an asphalt or metal flashing or interconnecting asphalt shingles.

chapter 13

Now Come the Hard Parts

Eaves, Rakes, and Trim

So far the construction of the house has been relatively easy, even for the amateur, especially if he's worked with a pro.

Ah, but now come some tricky details that might throw the amateur without help. They involve such things as eaves, boxed cornices, facias, soffits, and rakes, plus simpler things such as general trim and corner boards.

Exterior trim should be of a good grade of wood, especially where it is subject to moisture. Woods best resistant to moisture are cedar, cypress, and redwood. Where other species must be used, such as pine, they should be dipped in a wood preservative with a copper or pentachlorophenol base. This is especially important concerning ends of boards, where moisture can penetrate the end grain easily. Most trim will be painted, and that protects it. To do the best job, dip the trim, then paint it on both sides, edges, and ends with an exterior primer, all after making all necessary cuts and before installing it.

It is also a good idea to get the clear grade of wood; that is, without knots, because the knots must be sealed with two coats of shellac or other kind of stain killer so they won't bleed through the paint. Even with the sealer, there's no guarantee against bleeding, and there is nothing more frustrating than brown spots showing through almost any color where there are knots. When the trim is stained, it is not necessary to seal the knots. And when it is stained, the sealing with exterior primer need be only where the wood surface is hidden.

Clear-grade wood is very expensive. When wide areas must be spanned, ⅜-inch exterior plywood (one good side) can be used, and it will take paint very well.

Nails should be rust-resistant (galvanized, aluminum, stainless steel). Large-headed nails will hold trim nicely, but to give a neat appearance, heads should be countersunk (set) and filled with putty. Glazing compound, which stays pliable, is my favorite "putty" for outside work. Then the trim can be painted smoothly. Since large heads are hard to countersink, siding nails or casing nails with smaller heads can be used. Finish nails, with almost no head, can also be used, but their holding power is less, and unless the countersunk heads are filled immediately, the wood swells and closes the holes, not enough to cover with paint but too much to fill properly with putty. On siding, ungalvanized steel nails are sometimes used.

They will corrode a little and stain the siding, which has been stained or left to weather, just a bit, giving a rustic look.

As for the trim, let's start with the cornice at the eaves. The cornice is the extension of the roof at the eaves and the structure, usually boxed, that connects the roof line to the side walls. Its face is called a facia, and its underside is called a soffit, which is supported by a frieze board (trim at the top of the siding) and a molding.

The width of the cornice is determined by the extension of the rafters and their pitch (Figure 25 E, F). The shallower the pitch, the more they can extend over the wall. If the pitch is fairly steep, then the soffit might extend below the top edge of the windows, which are usually topped by the frieze board.

If the rafters are cut flush (with no tail overhanging the side wall, Figure 25 A), the cornice is called a close cornice. This is easy to build but is skimpy, and will look that way. The skimpiness is disguised by installation of a box wood or aluminum gutter.

I prefer a narrow box cornice in which the rafter is cut with a tail, and its end cut plumb and level, so its tail can be covered with a facia board and its bottom covered with a soffit (Figure 25 E). The facia usually overhangs the soffit by half an inch. This cornice will extend usually 6 to 12 inches, depending on the pitch and width of the rafters. It's always a good idea to make the soffit in two pieces, leaving a ventilation strip 2 to 3 inches wide, depending on the width of the soffit. To do this, staple screening along the underside of the rafters and then nail a strip of ⅜-inch plywood or nominal 1-inch board at the back and front of the soffit.

When nailing finish plywood, use siding or shingle nails. It is difficult to countersink nails in plywood, and siding and shingle nails have small enough heads so they can be driven flush and painted.

Cornices 6 to 12 inches are enough to give protection to the sidewalls and to give the roof line a substantial look. A wider cornice, with lookouts, is usually installed in contemporary construction, and not only protects the siding but helps shade the wall and windows in summer, when the angle of the sun is high, and lets the sun in the windows in winter, when its angle is low.

A wide cornice is built the same way as a narrow one, except additional members, called lookouts, act as a base for the soffit (Figure 25 F). They can be 2 x 4s, and are toenailed to the sheathing and into the wall studs with three 10d nails and face-nailed to the rafter faces with three 12d nails. Sometimes a nominal 2-inch plank is nailed to connect the ends of the rafter to act as a nailing header for the facia.

Cover the cornice with the proper-sized facia and soffit, installing the screened ventilating strip described earlier (this page). With a wide cornice (up to 24 inches), ⅜-inch plywood is good to use on 16-inch centered rafters. Anything less will tend to bow and look terrible.

Sometimes, when the overhang is so great that a boxed cornice with lookouts will extend lower than the top of the window, a boxed cornice can be made without lookouts. There are two ways to do this.

One way is to cut the rafter end plumb and nail a facia to the ends. The soffit, with ventilating strip, is nailed directly to the rafters, so it slopes at the same angle as the rafters (Figure 25 D).

The other way is not to cut the rafter either plumb or level. Then, slant the facia in and the soffit up, following the contour of the uncut rafters (Figure 25 C).

Either of these techniques can be used in traditional or contemporary design. The uncut rafter is, of course, cheaper both in use of materials and labor.

Even cheaper is an open cornice, with everything the same as the cornice without lookouts except that the soffit is eliminated. This construction is particularly effective with contemporary design, with rafters set at centers wider than 16 inches. Still another variation is to eliminate the facia board, often done in cottage construction. In traditional design, it is cheap and looks it. In contemporary design, it is highly effective.

There's one problem with an open cornice. The sheathing and siding must extend up between the rafters, so a nailer must be installed on top of the wall top plate, and the sheathing and siding must be cut around the rafters, a ticklish and exacting job. To eliminate this fussy work, the space between the rafters can be fitted with screened and louvered ventilators. This solution requires nailers, too. If the space were left open, it would be very inviting to squirrels and other vermin.

If you think a cornice at the eaves is tricky consider the overhang at the gable ends or the rake. Man!

You can decide on three different kinds of rake overhang: close, medium, and large. The large overhang, using lookout rafters of 2 x 4s, is described in Chapter 11. The principle is the same in other projections: to extend the rake so it won't be flush with the gable wall.

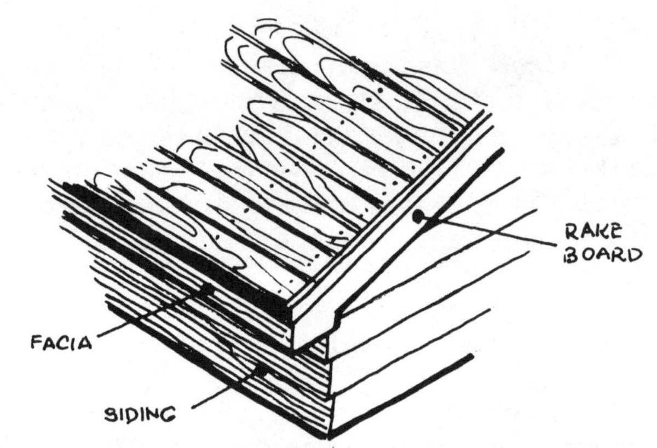

FIGURE 30. Close rake along gable end is easiest of rakes to build, requiring a simple 2-inch board to bring it out a little and a facia board. Cornice return forms the junction of rake and cornice (boxed eave).

A close rake (Figure 30) is simple: a nominal 2-inch plank is nailed along the edge of the rafter, under the extending roof sheathing. If the roof sheathing is flush with the siding, the plank is nailed flush with the roof sheathing. This plank is called a facia block, and a facia (rake) board is nailed onto it. Because this brings the facia out only 1½ inches, it does not need a soffit. To bring it out another 1½ inches, nail another nominal 2-inch plank to the first one. To bring it out 3 inches, nail short 2-inch blocks to the first, then another nominal 2-inch plank, then nail the facia board to it. Because of the short blocking, it will require a soffit. Traditional design incorporates this type of close rake, 1½ to 4½ inches, not counting the facia board.

If you want more overhang, first you must determine the extension and bring the roof sheathing out to this distance. For a rake overhang of 6 to 8 or 10 inches (Figure 31), short 2 x 4s or 2 x 6s are cut to the length desired and toenailed at 16-inch centers to the wall sheathing (through to the end rafter), with their ends connected by a facia board. This, plus the nailing of the overhanging roof sheathing to them, makes them stable. Nailing of a soffit supported by the frieze board and molding helps strengthen the lookouts. The molding and frieze board are installed on the sheathing, just as they are installed on the sheathing under the eave cornice.

A moderate overhang up to 20 inches is done the same way, with these differences: the lookouts, instead of being toenailed through the sheathing to the end rafters, are nailed directly to the nailing blocks and face-nailed with 16d nails through the sheathing and into the rafter. The lookouts are spaced on 16-inch centers, so the nailing blocks are 14½ inches long.

The other difference is that a fly rafter nominally 2 inches thick is nailed to the lookout ends with 16d nails. The fly rafter is connected to the ridgeboard, which is extended to the desired overhang when the roof rafters are installed. So you have to decide on the overhang when you install rafters. Facia and soffit, plus frieze board and molding, are installed as before. The molding simply makes a tight joint between frieze board and soffit. If it is tight already, caulking can substitute for the molding.

Now that we have that complicated affair done, how is the corner connected where the eave cornice and rake cornice meet?

When you have a hip roof, with eaves on all four sides of the house, the corner poses no problem. But when you have a gable end, you need what is called a cornice or eave return (Figure 30). With a close rake, the rake

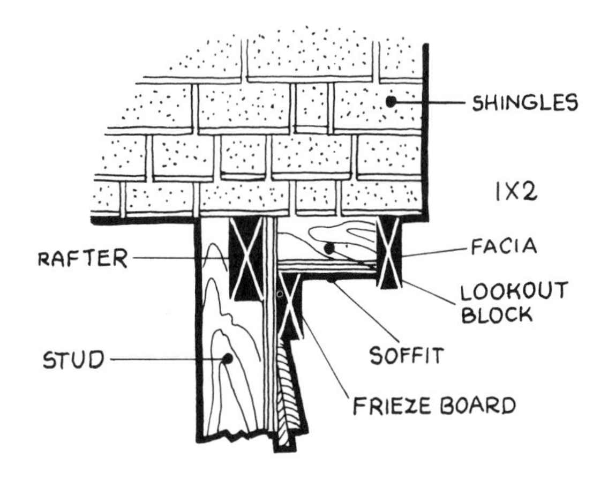

FIGURE 31. Medium rake, using 2 x 4 or larger lookouts, nailed through the sheathing and into the end rafter, to act as a support for the overhanging roof boards.

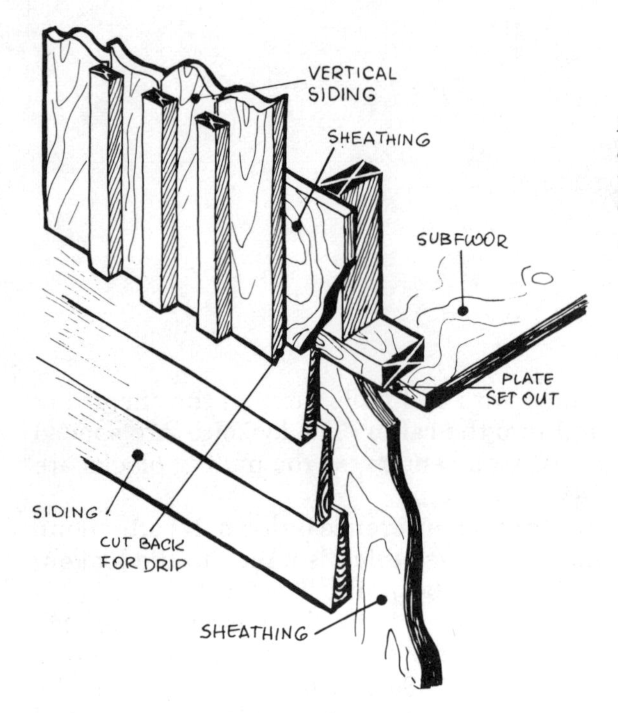

FIGURE 32. To change siding materials on a gable-end wall, set out the second floor plate so that the upper material will have its own overhang, just wide enough to act as a drip edge.

facia is cut at its lower end (or a separate piece installed) to match the eave cornice. This is usually necessary when the eave cornice is boxed.

Another way to do it, although less decorative, is when you have a slanting facia and slanting soffit: then the rake facia is cut at its end to match. When the eave facia is plumb, and the soffit level, then the rake facia is cut plumb, just as the rafters are cut plumb.

Finally, when both eave and rake cornices are boxed, the rake facia must match the contours of the eave cornice and be boxed just as the other two members are boxed.

Basically, it is more important in our opinion to have a medium eave overhang than rake overhang, because more rainwater flows over the eaves than the gable end.

Corner boards, an important part of exterior trim, are described in Chapter 16.

One more thing about gable ends. You may like the idea of the entire gable, that triangular wall from the peak of the roof to the top of the wall at the roof line, extending a little further out than the main wall below it (Figure 32). This gives more detail to the wall, and gives the chance of a change of siding from, say, horizontal clapboards or shingles at the bottom part to vertical siding at the top. Transition can also be made from brick to wood, and avoids cutting of bricks to follow the roof pitch.

An overhanging gable is built by installing 2 x 6s as gable wall studs instead of 2 x 4s, allowing them to overhang the wall by 2 inches. The upper siding then overlaps the lower siding by the same amount. Of course, the bottom ends of the overhanging studs must be covered by a nominal 1-inch board or by ⅜- or ½-inch plywood. It's a matter of design, but relieves the monotony of a large gable end, particularly if the attic or third floor does not have windows.

If you like the idea of a change of siding at the gable triangle, and can't or don't want to make the upper wall extend over the lower wall, it can be done this way: install a metal flashing in an L shape over the bottom siding, with the wide edge of the L extending under the upper siding. Support the drip edge (short edge of the L) by nailing molding at the top of the lower siding.

chapter 14

The Icing on the Cake

Roofing, Gutters, and Flashing

You now have the shell and inner "skin" of the house up, and next comes the icing on the cake: roofing and other waterproofing techniques, plus gutters, and in succeeding chapters, windows, and siding, which are further icing, all of which puts the house "to the weather": weatherproofs it.

Roofing not only must protect the house from weather, but it also must be durable, fire-resistant, and pleasing in appearance.

Fire-resistance is rated by Underwriters' Laboratories, Inc., "A," "B," and "C" in descending order of materials' resistance to flame applied to the surface and their ability to support combustion. Your building code will specify what roofing materials can be used, and within that specification, there are a lot of kinds of roofing, ranging from inexpensive to very expensive, or low durability to practically lifetime durability, from frequent maintenance needs to almost none. Generally, but not always, the more expensive the material, the longer it will last.

There are two types of roofing: multiple unit and membrane. Examples of multiple-unit roofing are shingles: asphalt, woods, asbestos cement, slate, and aluminum. They are used on relatively steep slopes (4-in-12 — 4 inches of rise to each 12 inches of run — and steeper), and protect the building by means of their water-shedding properties.

Examples of membrane roofing are asphalt roll roofing, built-up roofs (layers of felt applied with tar or asphalt and topped by mineral granules for long wear), and metal roofing of copper, terne, and aluminum. Membrane roofing resists penetration of water. Clay tile is in a class by itself, used on sloped roofs.

For most single houses with sloping roofs, asphalt or wood shingles are used, and that is what we'll talk about. Modern and other contempory designs sometimes use built-up roofs, and we'll add a word about that, plus roll roofing on very shallow pitches.

Asphalt shingles are inexpensive, easy to apply, and meet most fire codes. Wood shingles are more expensive, just as easy to apply, but may not meet all fire codes. Both materials are long-lasting, the asphalt lasting twenty years or so without maintenance and wood twenty years and up with low maintenance.

Asphalt shingles are the most commonly used, and come in various shapes, weights, and colors. Generally, a 235-pound weight of asphalt shingles (asphalt-impregnated felt

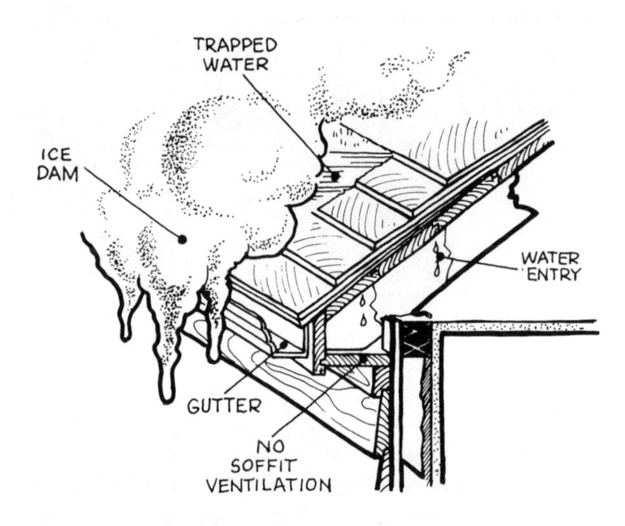

FIGURE 33. Causes of leaks from an ice dam: no flashing along eaves, no ventilation, and no insulation.

covered with mineral granules is accepted, but with changing standards, 250- and as high as 300-pound weight are being used more and more often. The number refers to the weight per square (100 square feet).

Color is mostly a matter of taste. Although light colors reflect the heat better than dark colors, if your attic is properly insulated and ventilated, color will have little actual effect on heating and cooling. Light roofs make a building look taller and larger than it is, while dark and black roofs keep it from soaring too high to the eye.

Although many codes don't even mention them, two things should be installed with a roof: flashing along the eaves to prevent leakage from ice dams (Figures 33, 34), and a drip edge to allow roofing to extend enough over the edge of the eaves or rake.

Ice dams occur in areas of moderate to heavy snow, and they really raise hell when weather is alternately warm and cold. Here is what happens: snow builds up along the eave edge. The more snow, the heavier it gets, creating ice along the roof. Then, when the weather warms or inadequate insulation and ventilation melt the ice, water forces its way through the cracks in the facia board and through the boxed cornice into the wall, ruining paint and interior finishes. Or, under pressure, this water actually percolates uphill and finds its way under the shingles, working beyond the upper edge of the shingles, soaking sheathing, and eventually dripping onto the attic floor, making more of a mess with insulation and ceilings. This problem is not to be ignored. Newspaper house columnists can vouch for the number of calls they receive from people in despair over ice dam damage.

Here's the solution: if you have ever traveled in extreme northern parts of the country, especially New England, you might have noticed houses with a couple of feet of metal flashing along the eaves, with the roof shingles starting not at the eave but a couple of feet up. The Yankees have some smarts, for this prevents water backing uphill and

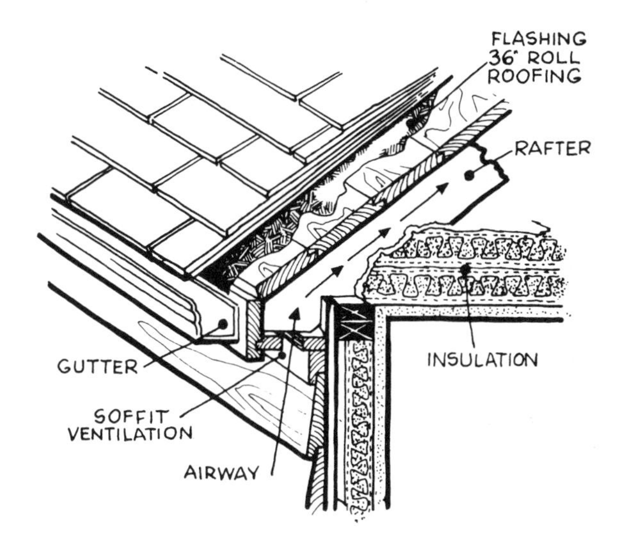

FIGURE 34. Roll roofing flashing along eaves, proper ventilation, and plenty of insulation are techniques not to prevent ice dams (snow may tend to build up to form a dam no matter what you do), but rather to prevent damage from water percolating uphill under the pressure of an ice dam.

into the house from the pressure of ice dams.

The same effect can be achieved by applying a strip of roll roofing weighing 90 pounds to the square, or 45-pound saturated felt, along the eave. This does not stop an ice dam from forming, but it does prevent water from backing up under shingles. The roll roofing can be nailed and/or cemented with roofing cement. Proper insulation and ventilation (see Chapter 18) will also help keep ice dams from forming (Figures 33, 34).

Underlayment is required for roofing in certain construction, and may be required by code. Underlayment is 15-pound saturated felt, also called roofing felt, and known to all twelve-year-old boys as tarpaper. Requirements can get complicated. First, it is not required when double-covered shingles are applied on a pitch of 7-in-12 or more. Double-coverage shingles have an exposure of 5 inches; triple coverage is 4 inches. Double coverage means there are two layers of shingle on each course (row) of shingles. Triple coverage means three layers on each course. Underlayment is also not required when the triple coverage is on a pitch of 4-in-12.

Single underlayment (where strips of felt are headlapped — overlapped 2 inches) is required when double-coverage shingles are on a 4-in-12 pitch and triple-coverage shingles on a 3-in-12 pitch.

Double underlayment (where strips are overlapped 19 inches) is required when double- or triple-coverage shingles are on a 2-in-12 pitch. Of course, roll roofing is a better roofing material for pitches of less than 4-in-12 because there is less chance of wind-driven water going between courses.

The second essential for roofs is a metal drip edge (Figure 35), a folded piece of aluminum that fits along the eave and rake edge of the roof, sticking out over the facia or rake board to allow water to drip away from the trim. Its service is improved if a 1 x 2 board is nailed into the facia board at its top edge. The metal edge also provides a firm base for the shingles themselves.

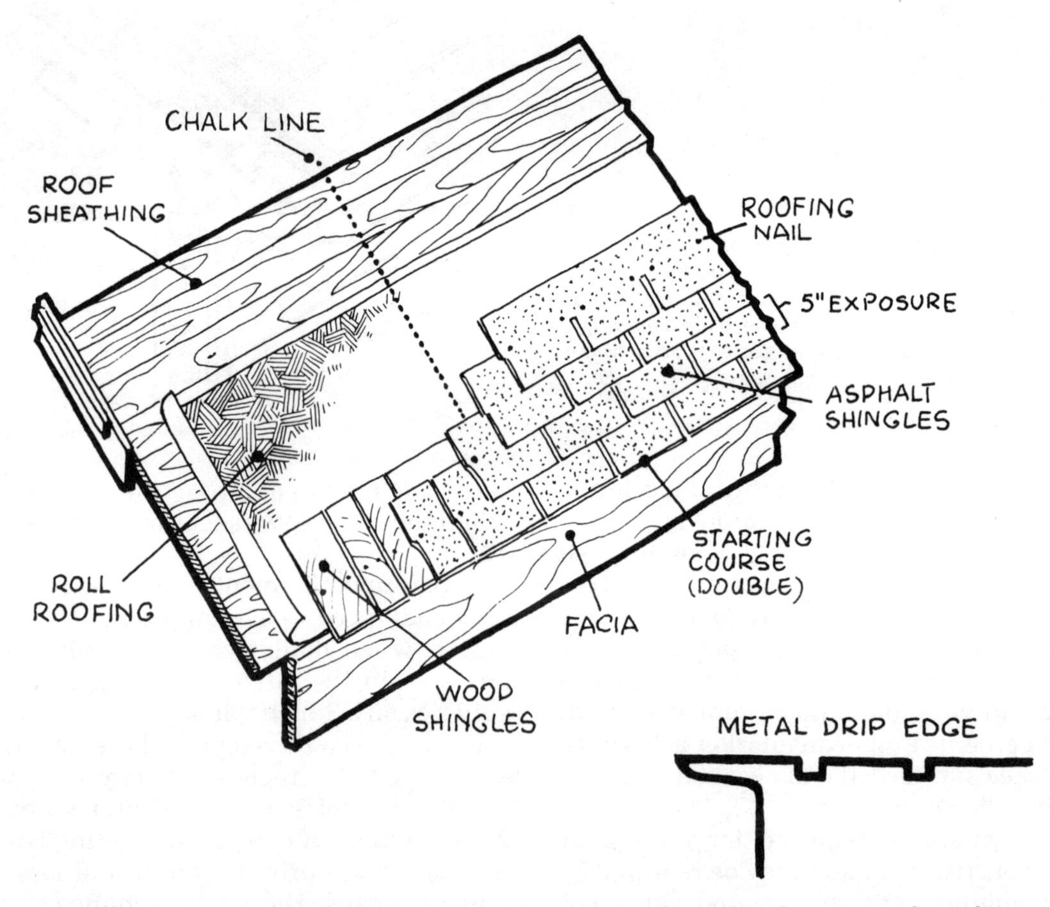

FIGURE 35. Roll roofing flashing goes on the roof edge. Wood shingles act as a drip edge, although a modern metal drip edge takes their place and will not rot out. Asphalt shingles can be of the "tab" type, with a groove every 12 inches to make them look like wood shingles, or the new type that have no grooves but rather scores to make them look like wood shingles.

Nail aluminum with aluminum nails. Galvanized nails will set up a galvanic action, causing corrosion. (Many roofers will dispute the damage that this will cause, noting that in many years, they've seen no damage.) Nail every 10 inches. The drip edge should be installed below underlayment at the eaves and above underlayment at the rake edge.

The drip edge takes the place of wood shingles placed at the eaves (Figure 35), as a first course, and sometimes at the rake, to allow a properly based shingle overhang. Now the wood shingles are eliminated, and it's a good thing because wood shingles under asphalt shingles last hardly longer than the asphalt shingles.

A common asphalt shingle is 12 inches deep and 36 inches wide, with two slots nearly half the depth of the shingle, dividing it into three equal tabs. Each end has a half slot so when it butts against the half slot of its neighbor the division makes it look like a whole slot. Some shingles are square on one edge with a whole slot at the other. The shingles have tabs on them to make them look like wood or slate shingles.

Relatively new, and growing in popularity, are shingles without tabs, but only shallow score marks to make them look like wood or slate. The slots are eliminated because, it has been discovered over the years, the roof wears mostly in the slots.

Either shingle is easy to apply (Figure 35): lay the first course topside toward the eave, then a regular course right side up directly on top of it. It is best, I feel, to overhang the drip edge by ½ inch along both eave and rake, although, as in many building techniques, there is not complete agreement over the ½-inch overhang. Some authorities recommend no overhang at the eave because the shingle will eventually sag over the drip edge. Instead they recommend that the shingles line up with the drip edge and that they be sealed with roofing cement to protect against water damage. However, along the rake there is a fairly unanimous consensus that the shingles should overhang by half an inch. If they do sag, you can always cut the edges off flush with the drip edge.

Use galvanized roofing nails 1 or 1¼ inches long. Drive four nails into each shingle, one above each slot, about ¾ inch above the slot. The same number of nails are driven into the slotless shingles, in the same approximate position. The nail head must be covered by the next course above.

Today's asphalt shingles have a strip or series of dabs of asphalt sealer on their backs to cement them down as a precaution against high winds.

Many asphalt shingles come with excellent installation instructions on the package label. Another good thing about these shingles is that they have small slits on each side, so when these slits are opened they act as a little hook so that the next course above can be automatically aligned. The slots are set for a 5-inch exposure, the most economical of installations.

When installing the starter course, cut the first shingle in half. Then, when the first course is laid directly on top of it, with a full shingle to begin the course, the joints will not match. If you laid all the courses beginning with full shingles, you would get the joints over each other, resulting not only in a leaky roof but a lousy-looking one.

When cutting shingles, use a utility knife. Cut on the reverse side, so the knife won't hit all those mineral granules, which are great dullers of blades. You can cut partway through the shingle and break it the rest of the way.

Now, lay the first course. The second course begins with a shingle cut exactly in half. The third course will start with a full shingle.

This will allow the slots of every other course, or the score marks of slotless shingles, to line up. Even if you use those little side slits that give automatic alignment, measure the exposure every now and then to make sure you're sticking to the right dimension. When you come to the opposite edge of the roof, you may find that you must cut off a shingle so that a slot is too close to the edge. Then the first shingle in each course must be cut to compensate for this.

Another technique with slotted shingles is to start the first course (on top of the starter course) with a full shingle, the second course with a third of a shingle, or 4 inches cut off, and the third course with two-thirds of a shingle, or 8 inches cut off. The fourth course starts again with a full shingle. This will make the slots of every fourth course line up, and may be considered more waterproof than the previous technique. This treatment is not necessary with slotless shingles.

Because asphalt shingles have such small exposure, and a roof line is usually pretty wide, the slightest variation in exposure at one end can result in disaster at the other, in the form of a badly slanting course, without the installer knowing it.

A chalk line snapped on the underlayment, say from the top of one shingle at each end, can do wonders in keeping the shingle lined up. And a vertical chalk line, from eave to peak, can help keep the tabs lined up (Figure 35).

When you're working on a roof, it's not a good idea to step back and admire or inspect your work. But every now and then go down the ladder and inspect the roof from the ground and "eyeball" the courses and slot lineups. And make the proper corrections and adjustments.

If you have to remove a course of shingles to make adjustments, bite the bullet and do it. It's your roof and you'll have to live with it. Any nail holes left should be sealed with roofing cement. Incidentally, 55 to 80 degrees is a good temperature range to work in. Lower or higher, both roofing felt and shingles can become tender and break easily. If the roofing cement is not pliable, keep the can in a bucket of very hot water. Professionals can work in lower and higher temperatures than you might because they're used to the vagaries of asphalt shingles.

At long last, we're up to the peak of the roof, having shingled up both slopes of the roof.

The most common covering for peak as well as hips is called the Boston ridge (Figure 36). We tried to look this term up in one dictionary and found twelve entries under "Boston," ranging from the city to Boston terrier, but no origin for Boston ridge. But at

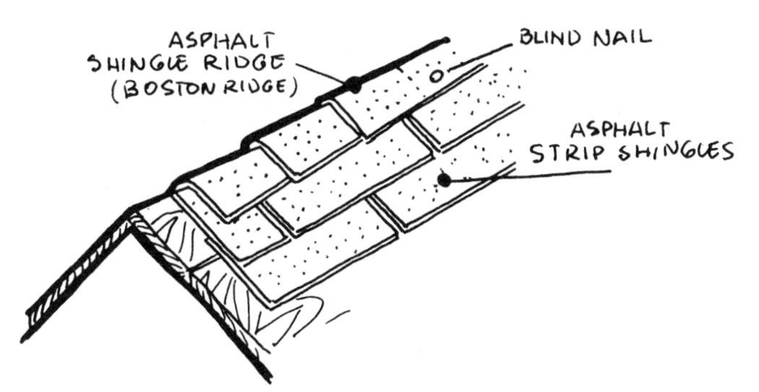

FIGURE 36. A Boston ridge is made by cutting a regular asphalt shingle into 12 by 12-inch squares and overlapping them away from prevailing winds. Nails are placed so the next succeeding shingle will conceal them.

least we can say what it is: Asphalt shingle squares are made by cutting a regular shingle into thirds, making them 1 foot square. They are folded equally to go over the ridge, each one overlapped 6 inches for double coverage, and each blind nailed on each side of the ridge, the nail heads covered by the next succeeding square.

If you can figure out the prevailing winds where you build, nail the squares so they are closed to the general wind direction. When you come to the end of the ridge, face-nail the last piece in two or four places and cover each nailhead with a dab of roofing cement. For areas of high wind and driving rain, a ribbon of cement under each lap also will help waterproof the ridge. Hips are done the same way as ridges, except the open end of each square should be facing the down slope.

Valleys are a little different (Figure 37). Here is where flashing applies. Where two roofs of equal slope meet, flashing is folded so that equal widths extend away from the valley. For valleys, use roll roofing. A good valley is made with an 18-inch strip of roll roofing facedown, with 9 inches extending on each side, and with a 36-inch strip face up on top of it, with 18 inches extending on each side.

Shingles are applied up to the valley, with a 4-inch exposure of the roll roofing at the top of the valley; that is, with 2 inches of roll roofing exposed on each side. The exposed roll roofing must widen toward the bottom of the valley at the rate of 1/8 inch per foot.

Try not to fold the roll roofing to conform to the wood roof; a sharp seam can weaken and break. Let the fold gently span the valley.

With metal valley flashing, folding is less

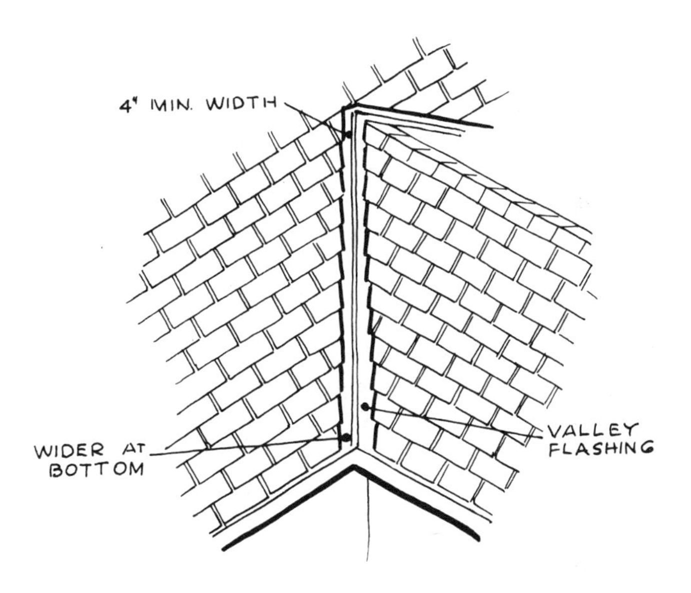

FIGURE 37. Valley flashing, where two sloping roofs intersect to form an "inside" corner. Flashing can be roll roofing or metal.

of a problem. You can use copper (very expensive but very long-lasting), aluminum (probably your best buy), galvanized steel, or lead. With metal flashing, the total width of the valley must be 12 inches for roof slopes of 7-in-12 or more, 18 inches for 4-in-12 to 7-in-12 pitches, and 24 inches for slopes less than 4-in-12.

Where a valley connects one steep roof with one shallow roof, metal flashing with a 1-inch standing seam in the middle of the valley must be used to prevent water rushing down the steeper slope from pushing up under the shingles of the adjacent, shallower slope. Install a ribbon of roofing cement under each edge of shingle, and use as wide a flashing as possible. Try to avoid piercing the flashing with the nails used for shingles.

Roll roofing flashing should be nailed with roofing nails at the edge. Metal flashing should be nailed with nails that match the metal used.

There are other forms of flashing that must be considered where the roof runs into a vertical wall or where the roof butts up against a chimney (Figure 38). These flashing areas are where most roof leaks occur, so the job must be done carefully.

Let's take the roof running into a vertical wall first. Any kind of underlayment should run up the wall at least 3 inches. Then, install step flashing, metal shingles that can be cut from any sheet metal (aluminum is still your best buy) and are at least 9 inches wide and 12 inches long. Even better is 12 x 12 inches.

Bend these shingles at a 90-degree angle and nail them so that half the width (minimum 4 inches) is on the roof (over the underlayment) and half (minimum 4 inches) goes up the wall sheathing. When the lowest one is installed, at the eave, the roofing shingle (doubled: starter and first course) goes over it, as close to the wall sheathing as practical. Then another piece of flashing is installed, overlapping the first by 3 inches. Then the second course of roofing shingles is installed, and so on until you get to the ridge. It helps to put a strip of roofing cement along the edge of the roof shingle where it lies on the flashing.

The idea of step flashing is to divert any water that gets under a shingle to the top of the shingle in the course below. This would be impossible with one long piece of flashing.

Siding is extended to within 2 inches of the flashing.

When the wall is brick, and around brick chimneys, the flashing is stepped in the normal manner. Where the roof descends to the high side of a chimney, a full length of flashing is required, with the roof sides of the folded L under the roof shingles. The side of the L that goes up the brick can be caulked with roofing cement.

Now, in addition, counter flashing is installed to cover the vertical side of the step flashing. This is usually a piece of preformed metal flashing with a lip on the upper edge, which looks like a stringer for a stairway. Mortar is raked out of a brick joint at the proper place at a depth of 1 inch, and the lip inserted and secured with lead wedges or other suitable clips. The joint is then remortared for a waterproof connection.

Sometimes, when a chimney is particularly wide, or the roof fairly steep, a saddle

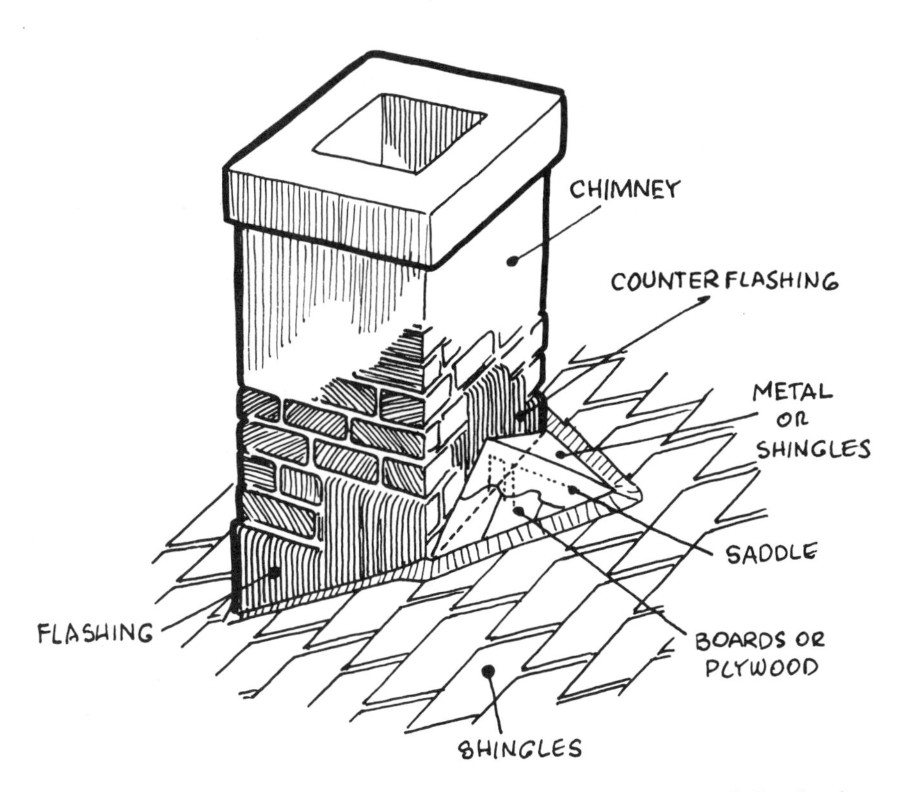

FIGURE 38. Flashing on a chimney, plus a saddle at the uphill side of the chimney, to keep water and snow from collecting on the high side of the chimney.

(Figure 38) is installed on the high side. This is simply a post next to the chimney (a 2 x 4) with a 2 x 4 ridge toenailed to the post and roof sheathing. It is covered in a double slope from the ridge to the roof with exterior plywood and then with a metal or shingle covering, with the cover's edge inserted under the shingles. The bigger the saddle, the more it is likely to be integrated into roof framing and covered with the same roofing material as the main roof.

Where the high edge of the roof abuts a vertical wall, such as on a shed roof, the flashing is considerably simpler. The underlayment is brought up the wall at least 4 inches. Then 12-inch-wide flashing is folded into two equal widths to follow the angle of the roof and wall. Roofing shingles are brought up to the edge of the wall and their upper edges generously daubed with roofing cement.

Then the flashing is installed on the wall sheathing with roofing nails, with the bottom part of the L shape allowed to sit right on top of the shingles. The area between shingles and flashing should be generously daubed with roofing cement. If the flashing doesn't set properly on the shingles, a sparing use of nails will help, but the nailheads must be daubed with roofing cement. Siding should go to within about 2 inches of the roof or more, if possible, to prevent snow or wind-driven rain from pushing up under the siding.

Incidentally, roofing cement not only comes in one- and five-gallon cans, but also in cartridges used in caulking guns, which can be very handy in roofing work.

Plumbing vent pipes and roof vents that must pierce the roof come equipped with collars that are their own flashing. These are installed with the collar under the shingles on the high side and over the shingles on the low side. Again, generous use of roofing cement is the best preventive against flashing failure.

Wood shingles for roofing are either red cedar or redwood, heartwood and edge-grained so they won't curl. White cedar is not recommended because the shingles tend to curl excessively. Top-grade shingles should be used.

Red cedar shingles come usually 18 inches long, about ¼ inch thick at the butt, tapering to ⅓₂ inch. They must be exposed 5 inches to the weather in a 5-in-12 slope and 3¾ inches in a 4-in-12 slope. Use of wood shingles (or asphalt, for that matter) in a slope of less than 4-in-12 is not recommended. For less than 4-in-12, use roll roofing. The wood shingles are usually resawn and rebutted, meaning they have 90-degree corners.

A ¼-inch butt shingle 24 inches long can be exposed 7½ inches on a 5-in-12 slope and greater, and 5¾ inches on 4-in-12 slopes.

A much longer-lasting shingle is a red cedar "shake," a much-misused word, but in this case meaning a shingle that is hand split on the exposed side and sawn on the bottom side, with butts of ½ to ¾ inch. They come in lengths of 18, 24, and 32 inches, and can be exposed, respectively, 7½, 10, and 13 inches.

An underlayment is not needed for the wood shingles. For the rough shakes, however, an underlayment is necessary.

Otherwise, the preparation is the same as for asphalt shingles. Spaced sheathing (Chapter 12) should not be used in areas of much snow.

Apply aluminum drip edge in the same manner as for asphalt shingles. A roll roofing strip along the eaves, to prevent ice dam seepage, is also applied.

Double the first course of wood shingles, which should extend ½ inch beyond the eave and rake drip edges. If you don't use a drip edge, extend the shingles over the eave line 1 inch and over the rake line ¾ inch.

Use two wood shingle nails in each shingle; 3d or 4d galvanized will do. When ½-inch plywood sheathing is used, use a ring-shanked galvanized nail. The nails are spaced ¾ inch from each edge and 1½ inches above the butt line of the next higher course.

Space all shingles ⅛ to ¼ inch to allow for expansion when they are wet. The joints between shingles must be 1½ inches from those in the course below. Never line up joints in one course with the joints in the second course above.

For shakes, longer galvanized nails are necessary and the procedure is the same, except the starter course can be the resawn and rebutted shingles. Also, underlayment, 30-pound felt, must be placed under each course of shakes. Some experts recommend 36-inch strips; 18 inches will do. Place a strip under the starter course and under the first course at the edge of the eave. Then place a strip about halfway up the shake so that its bottom edge is well underneath the butt of the next course.

Valleys are treated the same as with asphalt shingles. Choose the widest wood shingles and shakes, so the diagonal cut you must make won't cut across the entire shingle.

The ¼-inch shingles can be scribed with a utility knife. Thicker shakes should be sawed. If you have to split the ½- or ¼-inch shingles, use the knife. Thicker shakes can be split with a small ax.

For ridges and hips, a flashing of roll roofing or metal must be used. Shingles then can be applied like the Boston ridge described earlier for asphalt shingles, but with wood shingles, each pair is overlapped alternately. The alternate overlapping is to prevent the ridge from having the same joint exposed along its entire length. The open end of the shingles should be facing away from the prevailing weather.

A metal ridge also can be applied, with face-nailing covered with a dab of roofing cement. But a wood board ridge is better, at least for appearances. It is made by nailing a 1 x 5 to a 1 x 4 in an L shape, then nailing that along the ridge. The different widths of the boards compensate for the overlapping, making the exposure the same on each side of the ridge. This is less important than with corner boards, described in Chapter 16.

On roofs with a slope of less than 4-in-12, roll roofing is recommended. All preparation procedures are the same as for other roofing materials. An underlayment of 15- to 30-pound felt is necessary. Then, strips of roll roofing are applied, each overlapping the one below by half its width. Some roll roofing comes with the half lap already marked. The top half of each strip is nailed. The bottom half is secured by generous applications of roofing cement. This type of roof will not last as long as an asphalt or wood shingle one.

Other roofing materials, like slate and clay tiles, are very expensive, very heavy, and require special applications. They are not recommended for the amateur to install. The same goes for built-up roofs, which are for flat or very low-pitched roofs. If this type of roof has no slope at all, rain just builds up and sits there. But such roofs are designed to do just this. Sometimes there are drains in the roof to remove excess water. These roofs are made with 30-pound felt, the first layer laid dry (to keep tar from seeping through the roof sheathing) and secured with special roofing nails with 1-inch heads. The succeeding four or five layers of felt are separated by a layer of melted tar mopped on the surface. The final application of melted tar is sprinkled liberally with mineral granules (often called gravel) to make a good wearing surface. Because of the equipment needed for melting tar, this is not a job for the amateur.

Flashing on all types of roofs is similar, and since flashing is usually the weakest point of any type of roof, it must be done with special care and expertise.

Gutters (eaves troughs) and downspouts (leaders) are used when it is necessary to drain the roof runoff from the ground near the foundation. This is in turn necessary when excess water flow will seep through the foundation wall, or erosion or earth expansion will occur when runoff hits it.

The Colonials had a solution to this problem. They would dig their foundations shallow, perhaps 4 feet or less, and use the excavated earth to bank up against the foundations. Sometimes you may see an antique house that stands on a little hillock. Sometimes, for this reason, it stands higher than its neighbors. And in most cases you'll find the cellar of such a house drier than its neighbors. The piled-up fill allowed roof runoff to drain away from the house.

Like the Colonials, I think gutters are the invention of the devil. But in some cases you'll need them. These cases are when there is no runoff near the foundation, and when the earth is particularly absorbent. And narrow overhangs (6 inches or less) contribute to this problem.

Two other solutions are possible: An apron of concrete can be laid, 1 to 1½ feet wide, right next to the foundation, all around the house. This will divert runoff that much further away from the foundation. The other solution is to make the eave and rake overhang wide enough (12 to 24 inches) so the runoff will hit the ground far enough away from the foundation so it won't cause trouble. Of course, neither of these solutions is satisfactory if the earth does not allow runoff or absorbs it too fast. And with too fast a runoff you'll get erosion.

So it's a Hobson's choice.

So, if you have to, or want to go to gutters, here is what they're all about (Figure 34).

Wood gutters are the most traditional, usually made of decay-resistant wood like fir, cedar, or redwood. They are expensive. If they must be spliced, they should butt, be sealed with roofing cement, and be doweled or fitted with a spline, both expensive propositions. They can be painted the color of the trim, and can be treated inside with a wood preservative or boiled linseed oil, a yearly job, or with roofing cement, iffy because it can crack.

Metal gutters are either half round (usually made of galvanized steel and getting hard to find), or box type, of galvanized steel, aluminum, or copper, or stainless steel and copper alloys. Aluminum is your best bet for the price. It can be natural or painted to match your trim. It also comes ready-painted. Another good material is vinyl, usually white only, and more expensive than aluminum. Slip joint connectors are available for metal and vinyl gutters. Some gutters can be fabricated their whole length without seams.

What size gutter to use? The general rule is the measurement of the downspout. For each 100 square feet of roof area feeding the gutter, one square inch of downspout area is needed. And the gutter is approximately the same size as the downspout. Generally, the regular gutter that you can buy is adequate for an average one- or two-story house, particularly if downspouts are spaced within 40 feet of each other.

The longest run for one gutter is usually 30 to 40 feet. For a run of more than 30 feet, a downspout at each end is recommended, with the gutter sloping from the center to each side.

The slope of the gutters should be about 1⁄16 inch per foot. Modern hangers are concealed and hold well. Depending on the water load, the hangers can be spaced 2 feet and wider.

With wood gutters, ¾-inch spacer blocks are installed along the facia board, and the gutter nailed or screwed to these blocks with galvanized fasteners through the facia board, and into the ends of the rafters if possible.

One important thing to note when installing gutters is that the roof line, extended in an imaginary line over the gutter, should clear the outer edge of the gutter by at least ½ inch to prevent water from melting snow backing up under shingles, the familiar ice dam we discussed earlier in this chapter. To solve this, the gutter should be installed about 2 inches below the drip edge to allow water to overflow the ice-filled gutter without sitting on the roof edge and raising all sorts of trouble.

Downspouts come in the same materials as the gutters. Wood downspouts are rare, and usually found in antique, restored, or reproduction houses. Most popular are metal: plain round, corrugated round, plain rectangular, or corrugated rectangular. The corrugated types resist bursting when filled with frozen water. They are connected to the gutter by means of elbows that come in angles of 45, 60, 75, and 90 degrees. The downspouts are installed with strap hangers or fancier brackets, sometimes set off from the siding by blocks or by fasteners themselves.

Downspouts are joined with slip connections, to allow movement during temperature changes. You can also connect them with pop rivets or screws. When connecting downspouts, install the upper section inside the lower section, to prevent downflowing water from spilling out at the joint.

The lower end of the downspout, where the roof water collects in a big, steady flow, must be diverted. It can be done by an elbow attached to a splash block, which spreads the water widely and evenly over the ground. Or it can be diverted to a storm sewer, if you're lucky enough to have one. Never connect downspouts to a sanitary sewer.

Finally, it can be diverted to a dry well, which is simply a hole in the ground, at least 10 feet away from the foundation, and filled with coarse gravel (4- or 5-inch stones) or rocks collected from your property. The dry well will work well if the soil is permeable.

Some old roof designs make the gutter part of the roof itself, but this should be avoided.

It's a good idea to install mesh baskets at the top of each downspout, to prevent clogging. The gutters must be cleaned regularly. Some builders install a leaf guard, a mesh type material spanning the opening of the gutter. This helps to an extent, but such a guard also has to be cleaned regularly.

Sometimes, in areas of maple trees, the winged maple seeds get stuck, head down, in the mesh, and when enough of them collect, the gutter looks as if it has grown whiskers. Then the seeds have to be removed, one by one.

When you're cleaning your gutters a year or two after the house is completed, and see quite a few of those mineral granules that surface the asphalt shingles now in the gutter, don't be alarmed. This is a normal process and after a while it will stop. Even if it doesn't, the erosion of the shingles is one of the ways they wear.

chapter 15

They're the Lights of Your Life

Windows and Exterior Doors

In Chapter 9, we suggested that you decide on the size and shape of windows and doors in the house you're building, because rough openings (stud spacing) must be the right size to accommodate such windows and doors. A rough opening is usually half an inch wider and higher than the outside jamb of a ready-made window.

In most places in the country, you can buy ready-made window units, called setup windows. They include sash (the frame holding the glass), jamb (side and top members into which the sash fits), casing (outside frame), and sill (bottom member), plus essential parts such as stops, parting bead, weatherstripping, spring or weight-loaded mechanisms for keeping the sash in a preset position, and hardware. Sometimes storm sash and screens are included.

The units are made of pressure treated lumber (pine), which means they are treated under pressure with a wood preservative to resist water and decay (Figure 39).

The glass is single thickness or insulating, the latter being two pieces of glass welded or sealed together with ¼ or ½ inch of dead-air space, which gives the sash their insulating value. In very cold or hot climates, insulating glass plus storm sash can ease the heating and cooling bill.

If you buy sash and materials for making your own windows, be prepared for a lot of specialized work involving the making of rabbets, grooves, dadoes, and plows. These will be described briefly, but we recommend buying setup window and door units.

Thousands of words have been written about windows, their function, and their style, so we will try to keep the theory to a minimum, and mention the advantages and disadvantages to each type.

Windows serve three purposes: to let in light, to let in air, and to add to the good looks of a house.

To let in light: try to eliminate bright light and dark corners in a room. Provide glass areas of at least 20 percent of a room's floor area; codes generally require 10 percent, but more is better. It is difficult to put in too much glass area, given the insulating abilities of modern windows. If there is a lot of window space, heat penetration (or loss) can be controlled by the use of blinds and draperies.

Principal glass areas should be oriented to-

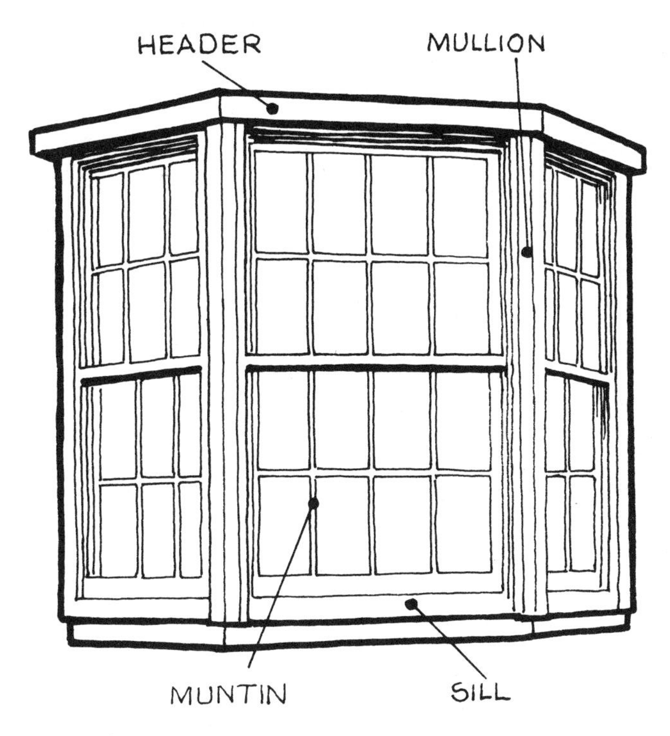

FIGURE 39. Window parts. Most window units, including a bow window like this one, come set up; complete with sash in jamb and casing, ready to insert in the rough frame opening.

ward the south in cold areas to take advantage of the heat of the sun. In warm climates, a northern orientation of glass is better to avoid the heat of the sun.

Place windows as high as possible for more light penetration. Place windows in more than one wall. Screen only the parts of windows that will be open for ventilation. A full screen can cut available light by 50 percent; half screens absorb only about 15 percent, according to one study.

Avoid coverings on the inside of the windows that partially block the windows in their open position.

To let in air: while air conditioning has reduced the necessity of this window function, there are still a lot of houses without air conditioning, and plenty of areas in the country where it is not needed. As a matter of fact, perhaps we will once again return to windows as ventilators in the not-too-distant future if our fuel situation gets worse without an alternate fuel source.

Locate windows to take advantage of the prevailing winds, if any. Avoid blocking the windows with shrubbery and trees, although these may serve more important functions, such as blocking the sun when necessary and blocking sound transmission.

Provide ventilation openings of at least 10 percent of the floor area of a room. If you have followed the rule for light (20 percent), half of each window, half the total area of the windows, will be available for ventilation, just 10 percent.

Locate windows for cross-ventilation, and for air movement within the level of occupants standing and sitting. High window openings do not provide such ventilation, although the design of the window, plus venetian blinds, can divert flowing air to the proper levels. Avoid corner windows, which ventilate a very small part of the room.

To add to the house's good looks: window design and location will accomplish this. These factors also determine the see-through-ability of windows, and the privacy they provide. A picture window may be of good design, but it is useless if what you see through it is not a "picture." And if privacy is not adequately considered, you may find yourself being the "picture" in that picture window as seen from the outside. House design is a compromise, and windows are among the most important design factors of a house, so the compromises must be weighed carefully.

What kind of window should be installed? It's a matter of taste, and the design of the house will help dictate what kind of window to use.

Traditional styles dictate traditional windows. Contemporary styles dictate contemporary windows. And so on and on.

A Colonial or Cape Cod house looks best with multi-lighted windows. A "light" is simply a pane of glass designed to admit light. A Victorian house needs very large windows, with good-sized lights. A modern ranch may take horizontal lights (three in each sash) to accent its broad, low-lying design. A modern house can use any type of window, but usually in an unusual shape, with lavish use of skylights, high windows, and window walls. Most modern houses today are built with a first-floor-level deck with sliding glass doors as access.

Modern windows often have one pane of glass in each sash, which makes painting and washing easy. To simulate the multi-lighted effect, a special grid of wood or rigid vinyl snaps into each frame, dividing the glass into several lights. It snaps out for maintenance. Grids are available for virtually every type of window made today.

For best service, sash, jamb and frame should be wood, and preservative-treated. Metal, usually steel windows (casements), were popular in the 1920s, but these have proved disastrous because metal is a very poor insulator, and when it is cold outside, the inside of the metal sash is nearly as cold, and the warm moist air in the house will condense on the cold frame, causing torrents of water to ruin the stool (inside sill).

Some modern replacement windows are made of aluminum, which is equally disastrous unless the outside and inside sections have insulation (urethane) sandwiched between them.

Now for the types, which can be installed singly or ganged or stacked two or more together:

1. Sliding

Double hung: a double-hung window consists of two sash that slide vertically, the inner sash in front of the outer sash.

To keep the window in position, the sash are supported at each side by springs, either concealed in the jamb or connected by a small cable to a concealed spring. Some double-hung sash can be removed easily. The sash weights in old-fashioned windows, while very effective, left an uninsulated pocket on each side of the window, and are no longer made, although you can buy sash as replacements for such windows.

Advantages: they are simple, widely available, and economical. Open, they don't stick in or out of the room. Sash tend not to distort over years of use because they are held in grooves at side, top, and bottom. A lock at the check rails helps to make the joint windtight.

Disadvantages: when open, they don't provide a baffle or scoop to bring cooling air into the room. They are awkward to open or close over a kitchen sink or high piece of furniture. They are poor protection from rain while open. Each sash, when open, covers its neighbor.

Horizontal sliding: regular windows slide readily in metal or plastic tracks. Sliding doors must have rollers; an overhead track is also good.

A sliding sash will stay in position where it is open. Like a double-hung sash, it is complicated to lock in an open position.

Advantages and disadvantages: generally the same as for double-hung windows.

2. Outswinging

a. Casement: sash hinged at the side to swing out. Inside swinging casements are not recommended because they are poor ventilators, interfere with drapery and shades, and limit the use of furniture near them.

Windows open with a crank: make sure the crank operates the entire window and not just the bottom rail.

Advantages: open, they will catch breezes and direct them into the room. They can be sealed and locked easily. They are easy to wash on each side, if hinges are set to allow arm space between frame and sash.

Disadvantages: storm and screen sash must be on the inside, or storm panels secured directly to the sash. If a crank is set into the frame, such storm and screen sash must be openable or provided with an access panel. Modern casements usually have the crank set outside the frame of the sash. Casements, when open, are an obstruction on the outside, so avoid placing walks and play areas near them. There is no protection against rain when this type of window is open.

b. Awning: sash hinged at the top, and opening outward. Contained in a single unit or stacked. The sloping sash in an open position looks like an awning.

They must be supported, when open, on both sides. This is usually done with a dual arm on the controls.

Advantages: same as for casements, and in addition, provide some protection when open against rain. Wind-driven rain is virtually impossible to keep out regardless of the type of window.

Disadvantages: same as for casements. Unless open to their full position, the downward slant of the sash diverts breeze too high in the room. When open, dirt can collect on the outside surface.

3. Pivoting

Jalousie: a series of glass slats that open like an awning window, in unison, like a venetian blind, operated by a crank.

Advantages: if slats can be opened at a greater angle than 90 degrees, air can be diverted downward into the room. Slightly opened, some ventilation is possible during rain.

Disadvantages: many openings between

slats result in considerable air leakage. Modern, more expensive units are better than earlier models, but if slat openings are glass on glass, the closing is inadequately sealed.

4. Inswinging

a. Hopper: Bottom-hinged or hopper windows are usually used at the bottom of a fixed window, so they can be opened to divert air upward into a room, can be washed easily, and will not interfere with drapery. They are sometimes used for basement windows, but the top-hinged, inswinging windows are better for basements.

b. Top-hinged: for high ribbon windows as well as for basement windows, to direct air downward. They can be washed easily, and except in the basement provide privacy without draperies.

5. Fixed

Windows that do not move. Usually they are "picture" types, flanked by movable windows, or with hopper types along the bottom. Because they need no hardware and can be secured permanently, they can be any size, and can be equipped with storm sash or insulating glass.

We mentioned earlier that all windows need storm sash or insulating glass. The climate really doesn't matter because such additional insulating will reduce cooling as well as heating bills.

Anyway, setup windows come in two basic styles: one with a casing (outside frame) made of ¾ x 3¾–inch pine and similar to a Colonial casing. The other has a casing of 1⁵⁄₁₆ x 2–inch pine and is carved or molded. The latter is narrower, obviously, and gives a "cleaner" appearance. It is suitable for Colonial, traditional, and modern styles.

The jambs (the side and top members holding the sash in place) must fit into the rough opening, and the jamb depth must equal the thickness of the wall. For instance, if the wall is made of 2 x 4s (3½ inches wide), ½-inch plywood sheathing on the outside and ½-inch plasterboard on the inside, the wall would total 4½ inches, and the jamb must be that thick so the inside casing can be nailed to the edge of the jamb without a gap between casing and inside wall.

And if your exterior wall is boards (¾ inch) and you use plaster (¾ inch total, including lath), you're adding ½ inch to the wall thickness, so you'll need a jamb 5 inches deep.

Windows can be ordered almost any jamb thickness, but they must correspond to the thickness of the wall. When you order windows, you'll be amazed at the number of styles, sizes — and depths — available. Your problem in the choice will be solved, presumably, by the plans of the house.

Many setup windows are prime-painted, so they only need one or two finish coats. Others are vinyl-clad, with a thin layer of rigid vinyl on all exposed faces except the sash, which is factory finished.

Before the window is put into the rough opening, the sides, top, and bottom of the rough opening must be fitted with an 8- to-12-inch-wide strip of roofing felt, to assure wind and water tightness between outside casing and sheathing. Leave 12 inches or so of tail on the bottom of the side strips; the reason for this will be given in the chapter (16) on siding.

The setup window is then placed in the rough opening, and shimmed with small wood wedges along the bottom edge to make sure it is level. Low-grade shingles are best for this purpose because of their taper. If the space between bottom of window and bottom of frame is too wide for one shingle, insert two, each in the opposite direction. These, when they are driven in, will bring the window to the required level. The same is done on each side to make sure the window is plumb. The top space does not have to be shimmed because the side jambs usually extend beyond the height of the jamb, forming "ears" on the top.

Now, drive casing nails into the casing (we always knew there was a use for casing nails), directly into the side studs and header of the rough opening. Casing nails have a tapered head, larger than that on finishing nails and smaller than that on common nails, so they can be countersunk and filled with putty.

Vinyl-clad windows have a special rigid vinyl flange extending from the molded casing. The window should be nailed with 1½-inch galvanized roofing nails every 4 inches to avoid nailing through the rigid vinyl covering the casing. Predrilled holes in the flange serve as a nailing guide.

Setup windows also come with a vinyl or aluminum drip edge to guide water down the siding and beyond the top of the window casing. Sometimes a wood drip cap is nailed under the flashing on top of the casing. This simply brings the drip edge farther away from the face of the casing.

All setup windows are installed in the same way. Bay and bow windows are installed similarly, although their projections present special problems, like roofing at the top (usually aluminum or copper) and support at the bottom overhang, which are provided with the window. If the window is supported by an overhanging wall, there is no other support needed.

If you really like the idea of constructing your own windows, or they are not available, there are two ways to do it. You can buy ready-cut units of jamb, sill, and casing, and build your own setup window. Or you can install jambs and sill in the rough openings, with special aluminum or vinyl side members on which the double-hung sash will slide.

While weighted windows work very well, they are not available. The aluminum or vinyl side members contain springs that hold the sash down or up, and their tension can be adjusted. The side members contain their own parting beads and exterior and interior stops, members which do the holding in place.

The sill goes in first, at a 3-in-12 slope to shed water. Then jambs are installed, and they all must be dead perfect (level and plumb) and perfect in size and shape so that the sash will fit properly. Then outside casing is installed and you're home free.

Other types of windows are installed according to their function. As you can see,

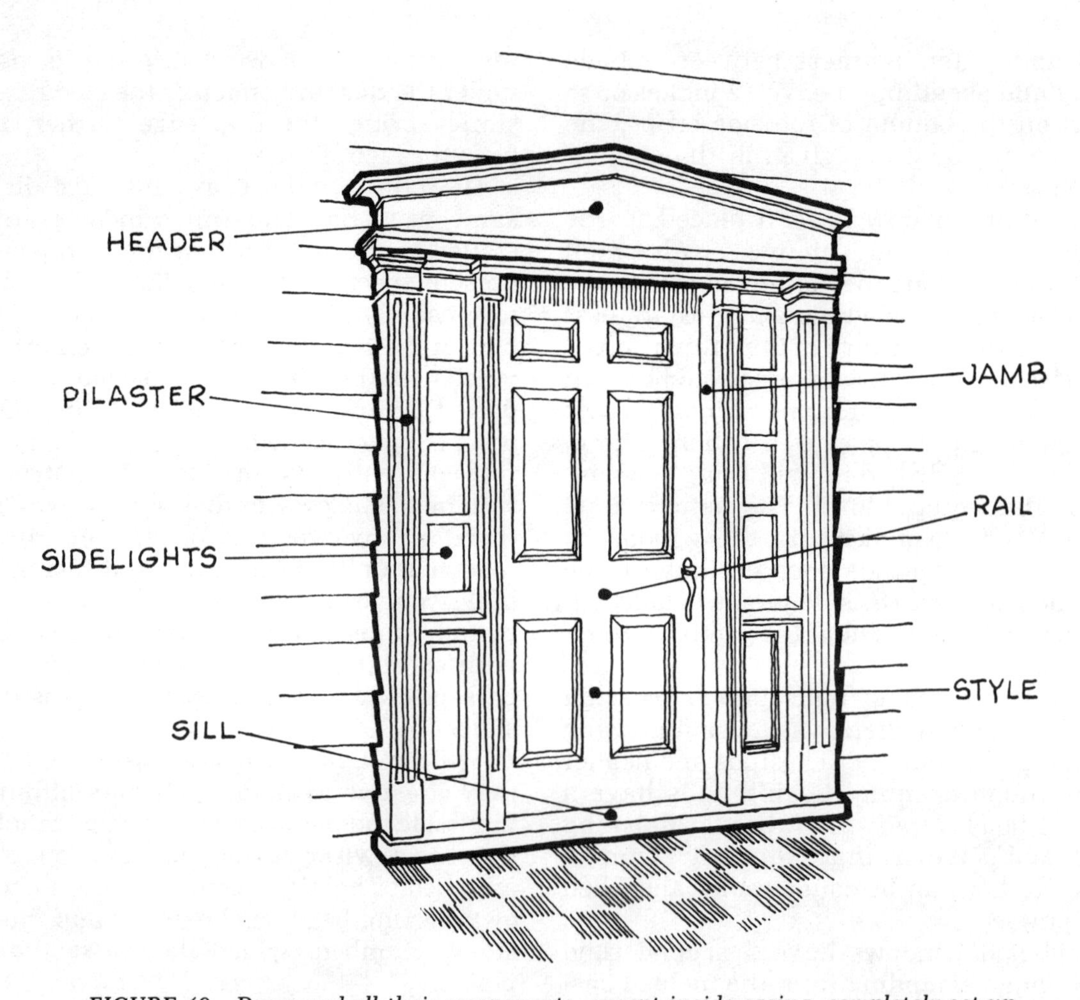

FIGURE 40. Doors and all their components, except inside casing, completely set up for insertion in a rough opening. This door is a "cross-and-bible" panel door, with rails and styles forming a cross on the upper section and rails, styles, and panels forming the open pages of a Bible on the lower section.

setup windows are certainly easier and faster to install than the do-it-yourself types. The latter might be less expensive, but if you plan to install say, twenty windows, you'll be on the window job for months. If for any reason you can't use setup windows, you should be able to find a cabinetmaking firm to build them out of component parts.

Now that windows are in, you can concentrate on doors (Figure 40).

Exterior doors are 1¾ inches thick, and come in many styles, from traditional paneled to modern flush. Most are made of pine, others of fir or of steel with an insulating core. Flush doors have a hardwood veneer on both sides, and the core should be solid, or at least filled with filler blocks for strength.

Doors come in standard heights: 6 feet 8 inches and 6 feet 6 inches. The front door is 32 to 36 inches wide, and service doors (side or rear) are 30 to 32 inches wide. The wider the door the better, to allow easy movement of furniture and appliances.

Doors are best purchased set up, with jamb, casing, trim, hinges, and locks installed. Set up doors can include sidelights (narrow windows on one or both sides), top lights, and even a fancy fan or sunrise light (pane) along the top. Trim can include head casings that are flat, equipped with a decorative gable, a broken pediment, or fan molding. Pilasters (half pillars set against the wall) come with many styles.

You must specify what size and style door you want so you can determine the rough opening.

The heavy-duty sill or threshold is 2 inches thick and made of oak. The ones made today are designed to set flat on the subfloor, with the top side sloped to shed water. Door jambs are cut to follow the contours of the sill. More expensive doors are weatherstripped with integral, interlocking flanges that really keep the weather out. Otherwise, weatherstripping must be installed on the outside of the jamb.

The finish floor (¾ or 25/32 inch) butts against the sill on the inside.

Door jambs are made of wood 1⅛ inches thick or thicker, so a rabbet (groove) can be made on the inner edge of the side and head jambs, into which the door will fit. This is different from window stops and the stops on interior doors, which are simply nailed or screwed onto the jamb. The rabbet makes the stop all one piece for extra weather protection.

Strips of roofing felt are installed at the sides and top of the rough opening, in the same manner as for windows. Nail the setup units in the rough opening by driving casing nails through the casing and sheathing into the side studs, countersink, and fill with putty. Also nail through the jambs into the side studs, countersink, and putty. Use 8d or 10d casing nails for fastening ¾-inch material, 12d casing nails for thicker material. Nail every 12 inches, and when nailing through the jamb, make sure the nails go through the shingle shims.

Incidentally, the shingle shims will be sticking out after they are driven in to make both windows and doors plumb and level. After nailing, you can break them off, but cutting them with a utility knife is neater and will reduce shim splitting.

When making a door frame plumb and

level, put a square into each corner to make sure the corners are square, and that you have not "racked" the frame (made it a parallelogram instead of a rectangle) while putting it in. To keep the side jambs from warping, put the shims close together.

A combination storm and screen door (wood or aluminum) can be installed in the section of a door unit where the exterior casing meets the jamb. The casing is set back from the jamb by ⅜ inch, so the door can be set in this opening. The storm door measurements must be exact. With a wood door, hinges designed for flush or offset doors are installed. To convert them from flush to offset, remove the pin, reverse one of the leaves, and replace the pin.

An aluminum door is supplied with its own aluminum jamb, which is lipped so that it can be inserted, door and all, into the opening provided.

We prefer wood storm and screen doors, even though they require painting and more maintenance than aluminum ones, which are at least getting better and better, and are a far cry from the early plain aluminum ones with all those sharp edges.

However, when it comes to storm and screen windows, we prefer aluminum because they are self-storing and require little maintenance. They come in anodized aluminum (some colors) and enameled colors as well as "mill finish" aluminum, which tends to corrode.

Aluminum combination windows come with two storm sash and a screen sash, each set in its own track; thus the name triple-track combination windows. In winter, the outside storm sash is at the top, the inside storm sash is at the bottom. In summer, the inside storm sash is raised to the top and the screen sash (innermost of the three) is lowered into position at the bottom half of the window.

Many windows, particularly casements, come with storm sash as a part of the sash itself, secured with clips. The screen is inside, attached to the jamb.

While many modern windows come with insulating glass, we would opt, if affordable, in areas of cold winters and hot summers, for triple glazing; that is, insulating glass windows plus storm windows. Such triple glazing, in combination with proper insulation, will reduce heating and cooling bills considerably.

You can have aluminum storms and screens custom installed. If you buy them yourself, you must get them in the right size. Before buying, measure the *inside* measurements of the exterior casing. In a window with wide casing, the storm is installed on the casing itself. With the window with narrow casing, the storm is installed on an offset blind stop. The latter requires exact measurements; there is a little more tolerance in size requirements on the windows with wide casings.

Combination windows come with clips that should be kept on until the window is installed because they hold the window rigid. If there are no predrilled holes in the flange, drill ⅛-inch holes every 6 inches.

Set the combination storm and screen window in the window frame to test for size. Then install the aluminum sill, a narrow strip with a U-shaped slot at the upper edge and a lip on the lower edge, and cut to the width of the combination frame. Then put a bead of caulking compound (use a caulking gun) around the inside of the frame sides and top, and place the window in the frame with its bottom edge in the U-shaped slot. Drive aluminum Phillips-head screws in the precut holes.

If the window does not have clips, leave one storm sash in the frame and center it, to make sure you don't squeeze the frame or open it up too much as you're installing it. In either case, the sash fit would be horrible.

The reason we dwell on do-it-yourself aluminum combination window installation is economy. There are dealers around (you might have to look pretty hard) who deal in missized combination windows, and others who sell them at special sales once or twice a year. Obviously, a missized window (one that is the wrong size for a particular job) is not worth much to the installer. So they are sold at very low prices, usually one-third to one-quarter the regular price. So if you're looking for ways to save, this is a good one. Just find the windows that will fit your window openings, and you've saved plenty.

One thing about the theory of storm windows. They are not designed to be airtight. The caulking is to prevent rain from entering. There are usually weep holes in the aluminum sill, and if not, you should drill three ⅛-inch holes in the sill. This will not be too much to prevent the air between window and storm from remaining still and therefore insulating, but will be enough for the windows to breathe, which isn't as farfetched as it sounds. When air heats up, it expands. When it cools, it contracts. When air expands in the space between inside sash and storm sash, it needs someplace to go. The ventilation holes are just such relief valves. This isn't too troublesome, because you won't get enough pressure from expanding air to do any damage. But when air cools it contracts, and it is possible for enough air contraction to occur, if more air doesn't come in to compensate, to cause a vacuum, and enough of a vacuum to break the glass.

chapter 16

Beauty Really Is Skin Deep

Siding

Once the exterior covering called siding is applied and stained, painted, or left to weather, the house is "to the weather," and will resist anything Maw Nature can throw at it. Not only is the house weather-resistant, but with lockable doors and windows, anything in the house is relatively safe.

You could even live in the house, except there are no facilities yet, and you have to have an occupancy permit in most communities to live in it legally.

Most siding is wood, which comes in many varieties. Other types are stucco, brick veneer, and the artificial sidings like aluminum and vinyl.

Wood is the most popular, least expensive, and easiest to install. Before you choose the type of wood, you have to decide whether to paint it, stain it, or let it weather naturally. Best woods to use in any case are cedar and redwood; cypress is good, too, but pretty hard to obtain. Most pines are good, but they don't weather well untreated or unfinished. Generally, the type of siding you buy will be available in one of the woods suggested. You really don't have much choice.

Any of the types can be used on traditional, contemporary, and "way out" styles. Vertical treatment is seen most commonly on contemporary and modern houses.

Clapboards are probably the most popular of the sidings (Figure 41). The original clapboards were ordinary square-cut boards, lapped over each other in succeeding rows. Later came bevel siding, which are boards tapering from ½ inch thick at the bottom to ¼ inch. This makes lapping them easier. Sometimes the butt is ¾ inch thick. The word "clapboard" does not come from someone named Clapp who invented them, but is a partial translation of Middle Dutch *clapholt: clappen,* to crack, split, akin to Old English *clappian,* to clap, and *holt,* board, wood.

A bevel siding with a shiplap at the bottom, allowing for automatic spacing, is called Dolly Varden, another example of the quaint language in the building industry. Dolly Varden, a character in Dickens's *Barnaby Rudge,* wore a nineteenth-century costume with a tight bodice and flowered skirt draped over a bright colored partially visible petticoat. It's also the name of a colorful trout. Crazy.

There is also horizontal drop siding, with tongued and grooved edges; some plain,

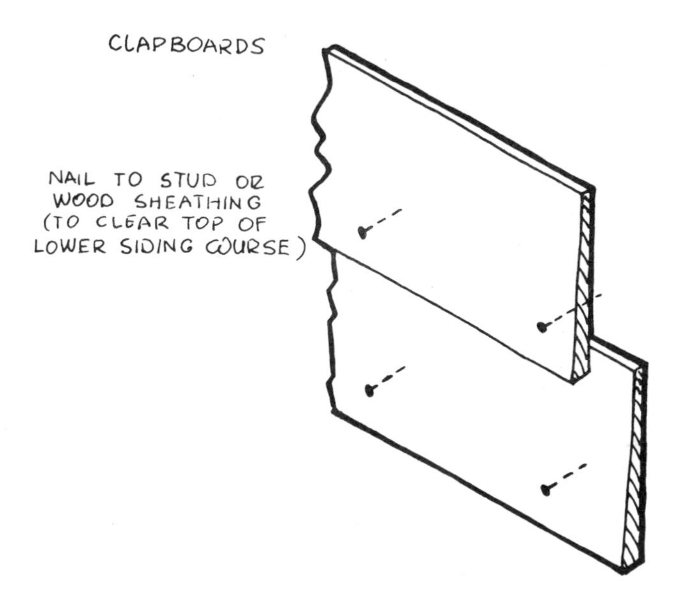

FIGURE 41. Clapboards should have at least a 1-inch overlap. Clear the top of the clapboard below when face-nailing; but be careful not to split the clapboard. Another technique is to pierce both clapboards, the one above and below, with the same nail, but this makes them more difficult to replace in case of damage.

others with molding. Some can be used vertically. Clapboard, bevel siding, and drop siding come in widths from a nominal 4 to 12 inches. The wider the exposure, the more contemporary the appearance.

When installing bevel siding, each board should lap a minimum of 1 inch. There is a continuing controversy over whether it should be applied over sheathing paper (a breathable, nonasphalt-saturated paper) or felt (a breathable, water-resistant but not vapor-resistant asphalt-impregnated paper). The community code might specify which. At any rate, either is acceptable, and while none is needed over exterior-grade plywood, something is needed over building boards.

Getting the boards to come out even is always a problem, when they touch the bottom of windows and tops of windows and doors. So here is how to do it: Suppose the height of the window (from sill to top of casing) is 57 inches, a nice odd number. If you are using 6-inch siding with a 4-inch exposure, you divide 4 into 57, to get 14¼ courses. Then divide 14 into 57 and you get 4.07 inches, which is less than 4⅒ inches. The fraction is so small that if you make the exposure 4 inches for all but one or two courses, which would be 4 and ⅛ inches, only you would know the difference.

But we still have to figure on the space from the bottom of the sill to the bottom of the first course. Say that is 35 inches in our hypothetical problem. Four inches into that goes 8¾ times. Divide 8 into 35 and you get just over 4⅓ inches. This is not compatible with the 4 plus inches of the siding beside the windows, so what do you do?

Two alternatives: pick an arbitrary exposure like 4 inches and notch the board to fit under the sill and over the window and door frames. It is best to try to get a full course width under the windows.

The other alternative meets the problem of many windows of different heights whose tops are even with each other but whose bottoms are not: Measure the height of the entire wall from the bottom of the first course to the bottom of the frieze board, the trim nailed to the sheathing just below the cornice. Say this height is 17 feet, or 204 inches. A 4-inch exposure goes into this 51 times total, or 51 courses. A 6-inch exposure goes into this 34 times. And so on. And notch the boards where necessary.

With drop panel siding, you can't adjust the exposure because it is tongued and grooved, so adjustments must be made on the course just under the frieze board. Of course, the frieze board can also be adjusted to prevent cutting the top piece of siding.

Horizontal siding is installed from the bottom up, and generally by face-nailing. If you don't want the nails to show, use siding nails, which are galvanized, small-headed nails whose heads can be countersunk and filled with putty and painted. Nonrusting nails like aluminum or stainless steel also can be used. They can be countersunk and filled or driven flush and painted.

If you plan to leave the siding exposed to the weather, or to stain it, the best method is to use aluminum or stainless steel nails and drive them flush with the wood. They'll show, but if you try countersinking and filling with putty, the filler will show more than the nail, unless you stain with a pigmented stain. (More about paint and stain in Chapter 26.)

For most bevel siding (½ inch), 6d nails are used. For ¾-inch siding, use 7 or 8d nails. If the siding is going over nonwood sheathing, use 9 or 10d nails and make sure the nail goes through siding, sheathing, and into each stud.

The first course should be blocked out with a board of similar thickness, to form a drip cap. Sometimes a 1 x 6 or 1 x 8 board is nailed at the bottom of the sheathing and is topped with a wooden or aluminum drip cap and painted the same color as the trim. This is called a water table and is generally used with Colonial design.

Nail each course, locating the nails far enough from the butt to miss the top of the next lowest course. The reason for this clearance is to allow movement when the siding shrinks and expands, especially important when using extra-wide boards.

Some builders recommend the nail go through both courses. The reason for this is added strength and to prevent the nail, if there is a gap underneath, from splitting the course being nailed.

Keep butt joints to a minimum. Use long boards under and over windows and over doors. Where you cannot avoid butts, make them over a stud, essential when nonwood sheathing is used. Stagger butt joints so they won't line up on adjoining courses.

Butt ends against window and door frames, and caulk around the frame first. This, plus the roofing felt strips beneath the casings, doubles protection against air filtration.

Paint or stain? Cedar clapboards, which are kiln- or air-dried, can be painted with oil or latex paints, using a primer before the finish coat. They also will take stain very well, but if you plan to stain, put the clapboards rough side out: they're reversible, and the rough side will take pigmented or clear stain better than the smooth side. Stain tends to be rustic or ultramodern.

You can also leave the smooth clapboards to weather, which cedar does to a dark gray or light brown. The weathering will be faster if your house is on the seashore or five to ten miles inland. For a rustic effect, use steel-cut nails to fasten the clapboards and do not countersink, but leave to rust. The corrosion will stain the boards in a unique way, particularly if the nails are lined up evenly on each course so that they make a straight vertical line.

Drop or novelty horizontal siding as well as lapped clapboards have a constant exposure surface ¾ inch narrower than their width. Drop or novelty siding is usually pine, and it does not weather very well, so should be painted or stained. Novelty siding is not considered very good style, although it is fully weatherproof. Drop and novelty siding should be installed with galvanized siding nails, long enough so that the length of their penetration into the stud is twice the thickness of the siding being held.

Panel siding is tongued and grooved, and the narrower widths can be blind-nailed through the tongue with a galvanized finishing nail. Wider boards should have an extra siding nail face-nailed toward the lower edge, countersunk, and filled with putty. Because of the necessity of filling countersunk nail heads, painting is most desirable.

Shingles (Figure 42) are another form of horizontal siding, and come in several species of wood, but red cedar and white cedar are the most common and most economical, as well as highly resistant to decay. When left to weather naturally, they will erode, instead of decaying, so that after twenty to forty years, they will wear down to a paper-thin covering. And don't think these old, thin, weathered shingles are not valuable; if you're in the right part of the country, and if you can find a buyer, they are worth big money, particularly if they are more than fifty years old.

At any rate, red cedar shingles come resawn and rebutted, which means their corners are 90 degrees, and need little splitting to get them to butt evenly against each other. However, you can install them loose, because chances are they will swell rather than shrink, when they are wet. Red cedar shingles are usually kiln- or air-dried.

Red cedar shingles come in 18- and 24-inch lengths, with butts ¼ to ½ inch thick. They can be installed from 5 to 12 inches exposed to the weather, depending on their length. The narrower the exposure the more watertight they'll be.

Red cedar shingles also come in the form of so-called shakes, with a grooved face, which are face-nailed with large exposures. They are usually dip-primed and ready for a finish coat.

Still other shingles are red cedar shakes, huge pieces of wood that have been split and resawn, with the smooth side designed to lie against the sheathing. These have butts up to ¾ inch thick, are laid with large exposures, and are sometimes installed with an uneven course line for a rustic effect.

Like clapboards, shingles with large exposures have a modern, contemporary look, while narrow exposures are more traditional. When shingles are tightly butted to each other vertically, they will give a strong shadow line and look a little like clapboards, especially if they are painted. When they are butted loosely, they'll give an individual shingle effect.

White cedar shingles come in 16-inch lengths, and are best laid up with a maximum of 6½ inches of exposure. They are not resawn and rebutted, so sometimes have to be shaved on their sides to line up properly. And they are not dried, but are "green" and have a high moisture content, so should be tightly butted against each other. They will shrink "in service," that is, installed, unless they are painted.

White cedar should be allowed to weather before painting or staining because of its high moisture content. If left unpainted or unstained, they will weather to an attractive

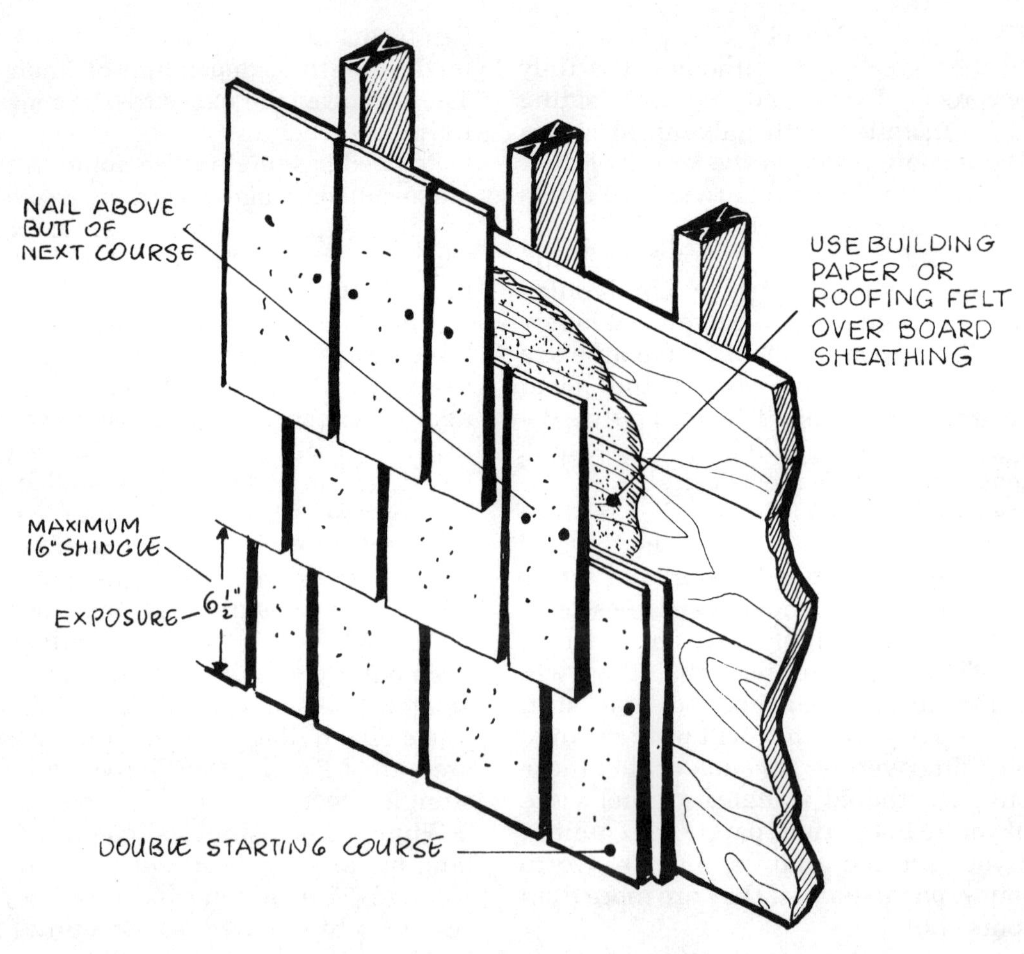

FIGURE 42. Wood shingles must be doubled on the first course (row). They can be tightly or loosely butted; white cedar shingles are green, so should be tightly butted to allow for shrinkage as they lose moisture. The main thing to remember is not to coincide vertical joints for at least three courses.

gray, the weathering effect enhanced by proximity to salt water.

When buying shingles, specify "extra" grade, which is free of knots. Some "clear" grades are not really clear. A wall of shingles can be marred when covered with knots. The knots also are difficult to seal when the siding is painted.

To install shingles, determine the amount of exposure. Then figure the height of the wall, the height of the windows, and the distance from the bottom of the first course to the bottom of the windowsills. Then determine if you can come to an exposure that will allow full course exposure over and under windows. A frieze board under the cornice is the only trim you have to worry about, except for corners, which will be discussed later in this chapter.

Double the first course, using second-grade shingles for the undercourse. Make sure the first course is level, because all other courses will be set according to the first. Use galvanized shingle nails, 4 to 6d. Shingles can be nailed directly through the sheathing, and need not be nailed to studs. If you use a fiberboard sheathing, nail a 1 x 2 or 1 x 3 furring strip to the studs at the bottom of the first course and along each succeeding course wherever the shingles are to be nailed.

The first course is doubled to create a slight drip edge, which allows water to drip away from the foundation. It also looks better; if you didn't double the first course, the wall would have a strange slanting-in look. Use second-grade shingles for the undercourse because they will be covered and are cheaper than extra grade shingles.

Small and medium exposures are blind-nailed; that is, about an inch above the bottom of the next succeeding course. Large exposures are face-nailed 2 inches from the bottom of each course.

Once the undercourse and first course are applied, use a 1 x 6 or similar straight board to butt the shingle bottoms against. Simply nail the board so that its upper edge provides the right exposure. To get it level, measure the exposure at each end and check each course with a 4-foot level. The longer the level, the more accurate it will be. Any holes made by nailing the board to the next lower course will not make any difference in weatherproofing. Use two nails for shingles up to 8 inches wide; three for shingles more than 8 inches.

Joints between shingles should be at least 1½ inches away from joints above and below. Avoid aligning joints two courses apart. The more staggered the joint, the more weatherproof the siding will be.

Vertical siding is another can of worms (Figure 43). It can be matched boards, or panels a nominal 1 x 6 or wider, the type that can also be applied horizontally; board and batten, which is spaced 1 x 6 boards or wider, with spaces covered by 1 x 2 or 1 x 3 battens; batten and board or reverse batten, just the opposite; and board on board, widely spaced boards 1 x 6 or wider, with spaces covered by similar boards.

These sidings are applied over roofing felt. When nailing over plywood or sheathing boards, use ring-shanked nails long enough to penetrate the sheathing for the pieces nailed directly to the sheathing. Use longer ring-shanked nails for the top boards or

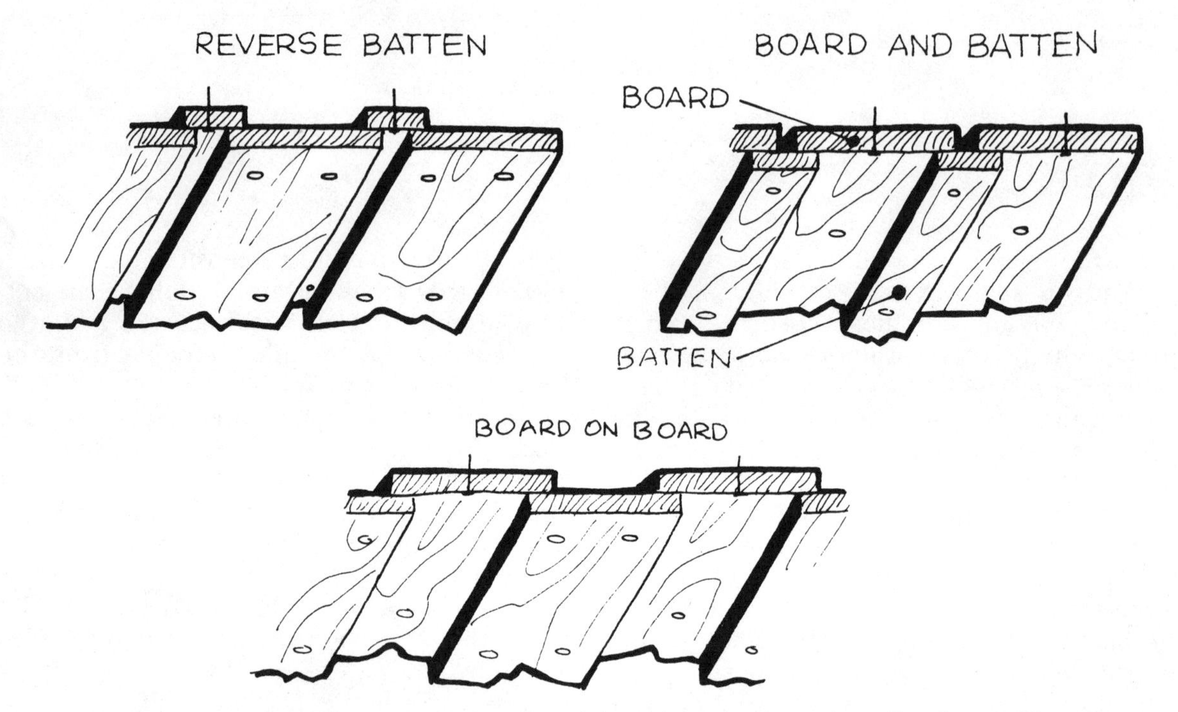

FIGURE 43. Types of vertical siding. When face-nailing countersink nailheads and fill with putty or glazing compound. Or nail them flush, but be sure to use nonrusting nails.

battens so that they will penetrate the sheathing.

If fiberboard sheathing is used, you can do it two ways: fasten 2 x 4 blocks between studs every 24 inches, and nail the vertical siding to these, so the nails penetrate 1½ inches into the blocks. Or nail 1 x 3 furring strips horizontally every 16 inches, nailing through the sheathing and into the studs. The latter technique is easier and takes less labor; and the ¾-inch space between furring strips will add to the insulating quality of the walls. Dead air is always an insulator.

Wider boards should be nailed near each edge, generally 16 inches apart. Battens can be nailed in a single row. When nailing outer boards, avoid nailing through the inner boards; always nail into the space, to allow for expansion and contraction. Use rust-resistant nails or countersink them before finishing. Vertical board and batten siding is available in 4 x 8–foot sheets of plywood ⅝ inch thick, with shiplapped edges, which can be nailed in the above manner. If the building code allows it, this type of siding can double as siding and sheathing at the same time.

With board and batten and other double-thick siding, door and window frames may have to be built up to extend beyond the siding. If the frames are flat, they can be built up by doubling the thickness of the frame, or nailing band molding along the outer edge of the top and side casings, mitering the corners.

Nonwood materials like aluminum and vinyl or mineral (the material that has replaced the old, familiar asbestos cement shingles) are sold with their own installation instructions, or are crew-installed. Stucco, a plaster made with Portland cement, is ap-

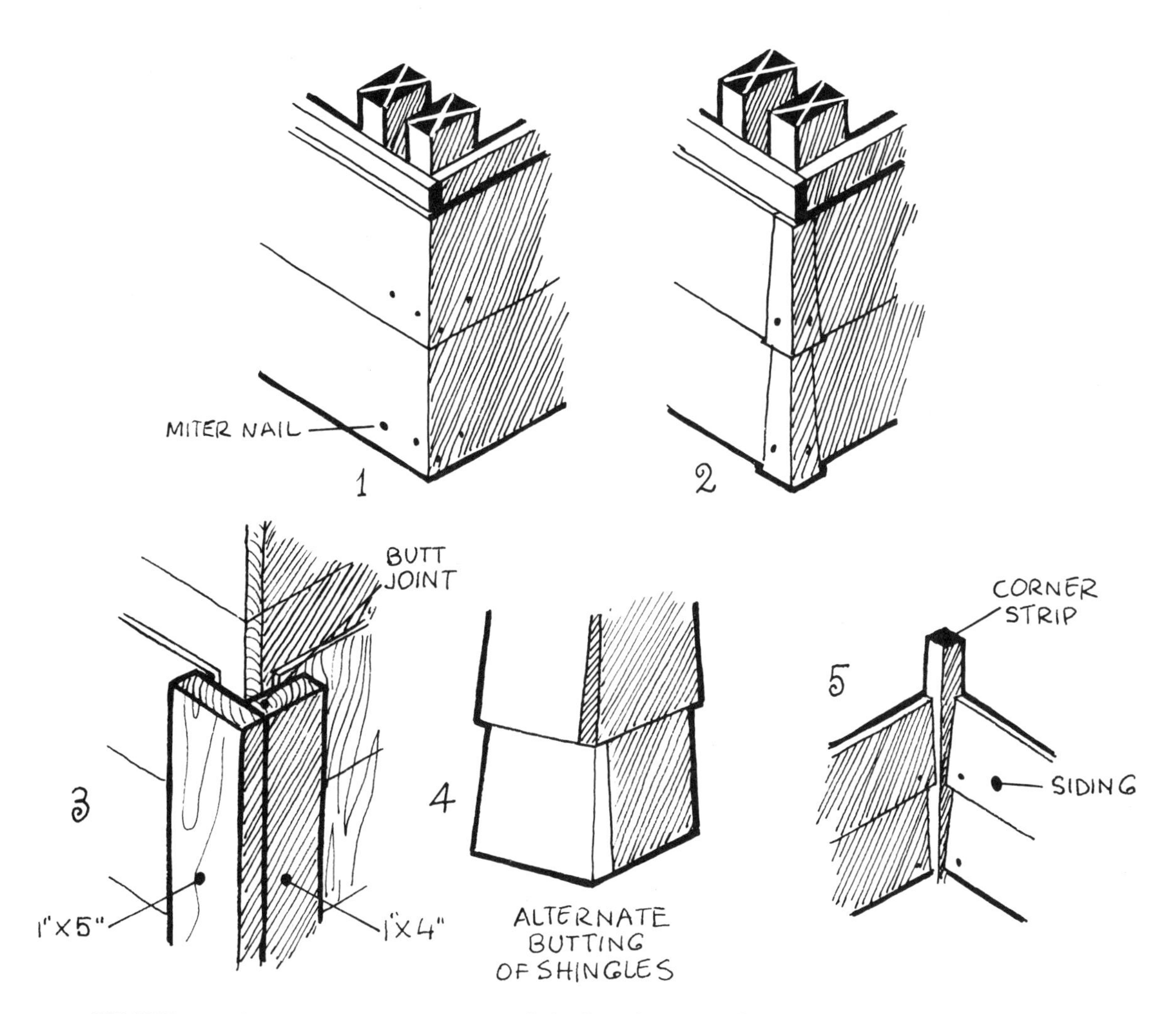

FIGURE 44. Corner treatment. 1. Mitered clapboards. 2. Metal corner covers (for clapboards and shingles). 3. Corner board (for clapboards and shingles: butt them up to the board). 4. Alternate overlapping of shingles. 5. Inside corner strip (1 x 1), with clapboards or shingles butted against it.

plied over an expanded metal lath, and except in some parts of the country, is not common. Our opinion is that wood is the best siding, other than brick.

Brick veneer is an excellent siding, but has special requirements — including a solid base on the foundation, window installation, and cornice treatment — that are beyond the scope of this book. But it will last as long as the house will stand, and needs only repointing (scraping out of crumbling mortar and application of new) every twenty to forty years. A brick house can be moved, but only very carefully. We've seen it done, and it's amazing to see how the house is held together with a network of steel rods and cables.

It's pretty simple to install wood siding, but corners can cause a man to pause (Figure 44). Here is the technique: on an outside corner, clapboards can be mitered to meet, but the work must be precise to be effective. Clapboards and shingles can be butted at the corners and covered with metal corners that fit over each course. Metal corners are good only when siding is painted.

Shingles can be overlapped at the corners, with the overlapping reversed for every other course. For both clapboards and shingles, corner boards are an excellent solution. They generally are of traditional design, but stained the same color as the siding, they can have a modern effect.

Corner-board dimensions depend on the size and height of the house; the higher the wall the wider the boards. If you choose a wide corner board, butt a nominal 1 x 6 or 1 x 8 against a 1 x 5 or 1 x 7 (the odd sizes might have to be cut to size), in the shape of an L. The edge of the smaller dimension is set against the flat side of the larger dimension. This way the two sides of the L will come out even, or nearly so. Nail cornerboards over a strip of roofing felt folded and nailed over the corner.

A traditional but elegant corner board is done this way: nail two boards of equal width right at the corner, so that their inside edges (touching the sheathing) touch at the corner, over a strip of roofing felt. Fill the miniature inside corner thus made with a ½- or ¾-inch quarter round, making a rounded, evenly sided corner board.

Butt all clapboards and shingles against the corner boards, which must be plumbed vertically to keep the house from looking crooked.

Inside corners are simpler. Over a strip of roofing felt, nail a plaster ground (¾ by ⅞ inch) or 1 x 1, and butt the clapboards or shingles against this.

chapter 17

Leave It to the Pros

Electricity, Plumbing, and Heating

In a finished house, we take for granted that lights turn on when switched, plugs work, the stereo sings (blares?), the toilets flush, the water flows, and we're reasonably warm. But when things go wrong, nothing seems right with the world.

So now is the time, when the house is to the weather and before insulation goes in an interior wall and ceiling surfaces go up, to install electrical wiring, outlet (plug) boxes, switches. Also, plumbing and other functions that require pipes or wires, such as burglar alarms, fire alarms, antennae for FM radio and TV, a vacuum system and anything else planned for the future.

It is not within the scope of this book to describe step-by-step methods of installing wiring or plumbing, but simply to plan for it. National, state, and community codes generally require such installations to be by a licensed expert, to be duly inspected and approved by the proper authorities.

One way to save money is to hire yourself out to the electrician or plumber as a "helper," if he is willing, to assist him in the rough installations. The expert then would make final connections.

Early in the construction of the house, the electric company must be consulted for connection of temporary service with the aid of an electrician. This service will run electric saws and other equipment that speeds and aids construction.

What capacity should a house have? For a house with up to 1000 square feet of floor area, 125 amperes is the minimum. For up to 2000 square feet (a large house), 200 amperes is minimum. These capacities are sufficient for lighting, portable appliances, and for equipment for which individual circuits are needed.

The capacities are also sufficient for electric heating and air conditioning of the individual room type. More capacity is required for electric heating of the central type, such as a heat pump or furnace or boiler that is fueled by electricity, or for central air conditioning.

At the service connection, where the electricity comes into the house from the street, there is a meter, usually on the outside of the house. The service from the street can be installed above or below ground, and the whole system is grounded by wrapping a ground wire around a steel rod stuck deeply in the ground or an underground water pipe. Inside, at a convenient location, usually in the basement or utility room, a distribution box contains as many circuit breakers (fuses in an older system) as there are circuits in the house, plus some spares. The circuits have capacities from 15 to 50 amperes. Lighting usually takes 15-amp circuits; light-duty appliances 20 amps, and heavy-duty appliances even more: electric dryer, 30 amps and electric range, 50 amps.

A fuse or circuit breaker is rated at 15 amps and up, and when a short circuit or other trouble occurs, the fuse blows or the breaker trips, cutting off the juice. But if a person causes the trouble, he can be electrocuted or badly burned before the juice stops.

Light-duty circuits are of 120 volts; heavier-duty ones like air conditioners, in addition to range and dryer, take 240 volts, with specially shaped outlets.

You should have enough circuits in your system so that all electrical functions, from the smallest light to the heaviest appliance, do not overload the system. For example, each appliance, such as portable kitchen items, disposer, compactor or dishwasher, operates on its own circuit, with nothing else connected to it. If other units are connected to such a circuit, overloaded wires could heat up and cause a fire, or at least continuous or excessive tripping of circuit breakers or blowing of fuses.

The wiring used in most areas of the house is No. 12, a fairly heavy-duty type. The smaller the number the heavier the wire. Some heavy-duty appliances, such as range, dryer, electric water heater, and electric heating, require wiring specified for their use.

Generally, wire is a three-wire 120/240-volt circuit, which means it is two 120-volt circuits using a common neutral wire. It's a combination wire, on which either 120 or 240 volts can be used, and is encased in a plastic sheathing, which can be simply strung around the house. It also has a ground wire. The wire should be made of copper. Some new wire is made of aluminum, but this kind has caused difficulties in connections at switch and outlet boxes, with incidents of dangerous overheating, and is not recommended.

If all this information sounds confusing, check with your electrician, who will make it clear, and if not, will make sure the job is done right. In fact, inspections should assure that the job is done right.

A fairly good-sized house can have up to forty circuits. The more circuits, the less likely that any one will be overloaded, and the less trouble there will be with overheated wires, tripped circuit breakers, and other problems, such as lights that won't light up to their proper level, motors that burn out because they're overloaded or underloaded, and TV pictures that get small when they shouldn't. A relatively new protection for the consumer called the ground fault interrupter system (GFI) has appeared, which gives maximum protection for people against shock and for appliances against breakdown. Certain appliances are protected by GFI.

A GFI is designed to detect and interrupt ground faults (leakage of current) between the line and neutral current in the circuit. A current imbalance as low as .005 amperes or

5 milliamperes (a milliampere is 1 thousandth of one ampere), will cause the GFI to trip.

It's not the voltage that causes the problems, but the current (amps) level being driven through a human body that causes burns or death. Anything over 100 milliamperes can cause death or burns under certain conditions (dampness, water, or contact with pipes). So appliances where water is a danger, plus certain outdoor connections, should be protected with GFI.

How many outlets should you have in each room? And how many switches? And how many ceiling fixtures? The questions can be endless.

The answers depend on your needs. Generally, no point along a wall in a room should be more than 6 feet from a receptacle. If one receptacle were on each wall, this requirment might be fulfilled, but if the receptacle were in the middle of the wall, it might be covered by a big piece of furniture and inconvenient to use. So, a receptacle every 6 feet is a good standard, or at least one near the ends of each wall so that plug outlets will be accessible.

Outlets should be 12 to 14 inches above the floor. In kitchens, bathrooms, and laundries, they should be located for easy access, either 36 inches from the floor or a few inches above the counter top. A dining room should have at least one high outlet so appliances can be used at the table. Kitchen outlets should be spaced much more closely than in other rooms for convenience.

Switches should be installed where they can operate ceiling or wall fixtures. Multiple switches should be used where there is more than one entrance to a room, and on stairways, allowing the light to be turned on or off at either entrance, or at the bottom or top of the stairs. They are set 48 inches from the floor.

In most rooms, outlets can be of the split control type: one outlet alive permanently; the others controlled by a wall switch so a room that has no ceiling fixture can be lit by a table lamp, for instance, that is plugged into an outlet controlled by a switch.

There are many types of outlets, but the main ones are 120 volt, indoor and outdoor. The outdoor outlet is weatherproof. And there are safety types that require the insertion of a plug and a twist to activate it.

Switches are of several types, too, but the most common type is silent. Others are waterproof, key operated, operated by closing or opening a door, time delay, and dimmer switches, for romantic dinners in the dining room.

Laying of cable is relatively easy in a house before insulation and wall surfaces are installed. The cable is installed with large staples. The only restriction, in some codes, is that wire cannot be strung between exposed floor or ceiling joists; people might hang things from it. To solve the problem, a 1 x 3 or similar-sized board must span the joists and the wire attached to it. Where a wire must be spliced, a junction box is required. It can be covered with a blank cover.

Switch and outlet boxes can be steel, plastic, or ceramic. Steel is your best bet. Such boxes must be installed so that the wall surface is flush with the front of the box. So, when boxes are installed, you must know how thick your wall surface (plasterboard, plaster, or paneling) will be.

Boxes can be installed midway between joists or studs by the use of special steel arms or on boards spanning the joists or studs. They also can be nailed directly to the joist or stud.

It's much easier also to take care of all your wiring needs before insulation and wall surfaces are installed. And it doesn't hurt to provide lighting in closets, hallways, and obscure places such as accessible but unused attics, basement corners, and stairways of all types and locations. Exterior applications consist of outlets in the back and front of the house, and installation of security lights, such as floodlights at the front, back, and sides of the house.

Unless you're content with a nice brass door knocker on front and back doors, install an entrance signal system. You can use a lightweight wire like No. 18, with a transformer, because the signal system uses less voltage than the house voltage, so the transformer reduces it. The signal should be located in one or more places where it can be heard, and a different signal should be used for front and back doors. A voice communication system can be installed, too, or at least wiring provided, even if you don't plan one at the moment. Such a system provides added safety, and when you want or can afford to install one, the wiring is already there. A whole intercom system can be wired now, and installed later.

At this point wiring for fire and burglar alarms is important. A fire alarm can be hooked up to both heat detectors (200 degrees) installed in every room and smoke detectors installed at the top of each stairwell, including the basement. An alarm bell would wake the occupants in case of danger.

The burglar system can be a pulsating siren activated by installing trip devices or weight-activated pads imbedded on the floor just inside doors and windows. The siren should be loud enough to scare away any burglars, who don't like sound. Both fire and burglar systems can use batteries in addition to the regular power, in case of power failure. Batteries should be inspected twice a year, and it's wise to replace them once a year whether they're working or not.

A television and FM antenna system can be installed when the walls are open, feeding outlets to rooms where you plan a TV or radio. The antennae themselves are best located in the attic to keep them from spoiling the outside contours of the house. Speaker wire can be strung from where the radio-phono-tape system is located, so you can put remote speakers where you want them. Telephone wire also can be strung, with the help of the telephone company, and plug-in outlets placed where desired, even if you don't plan to carry a phone from room to room.

Finally, a built-in central vacuum system can be installed, using flexible or rigid pipe and plug-in outlets that activate the vacuum cleaner in the basement. A 30-foot flexible pipe attached to a nozzle would allow a minimum number of outlets to the system. Again, it would be much cheaper to install

all this piping and outlets when the walls are open; then you can buy the actual vacuum cleaner when you're ready.

Suppose someday in the future you want to air-condition the house. You could put ducts in, which should be separate from a hot-air duct system because air conditioning ducts do their duty high in the wall so the cool air falls as it comes out of the outlet. This can be expensive, so you might want to consider an alternative: string the 120/240–volt wire to a strategic spot (under a window, for instance, or high on an outside wall, but located so it won't destroy the looks of the house), connect it to an outlet box, and put a blank cover on it. Then when you're ready for room air conditioners, just buy the units, put 'em in the window, take off the outlet cover and install the correct kind of receptacle, and turn on the cool.

Plumbing

Like the information on electricity, the information on plumbing in this chapter is designed as a guide, not as an installation manual.

The design of your house may or may not dictate economy, but it is economical to plan all your plumbing pipes and lines to be as close as possible to each other. For example, in a one-story house, the bathroom should back up to another bathroom or to the kitchen. This is not only to share the plumbing but also to share the space where pipes are installed. If the placements work out, you can also share the main drain line, which acts as a vent, too.

Basic plumbing is divided into three functions: supply (getting water into the house), fixtures (utilizing this water), and drainage (removing waste). Also basic is the requirement that drainpipes must have a vent. If there is no vent to relieve pressure built up by drainage of waste liquids, the plumbing will be plagued by strange noises and worse, siphoning of liquid from traps, thereby releasing sewer gases into the house by way of the fixtures.

Plumbing is of various types: cast iron, galvanized steel, galvanized iron, brass, copper, and plastic. Heavy drainpipes can be of cast iron or plastic, if plastic is allowed. Water supply and small drainage pipes are copper or plastic. Copper is probably the best material to use at the moment, although plastic will most likely be the plumbing of the future.

It is not within the scope of this book to explain how to install plumbing. Copper requires sweating of joints, which, simply, is a technique of soldering. Plastic pipes are fitted together dry, then when every pipe is in the right position, joints are secured with a special adhesive. Once they are secured, they are virtually impossible to take apart. Cast iron and galvanized steel are connected with special sleeves, compounds, and threaded connectors, and are even more difficult to work with than copper or plastic.

At any rate, there are certain techniques in installing plumbing in framed walls and floors. Because the pipes are of different sizes and are rigid, the problems are more complicated than with wiring. The vent pipes and toilet drains are the largest pipe that you will use, 3 inches in some cases, 4 inches in others. When 3-inch pipes are used, a 2 x 4 floor and top plate are permissible, but the top plates must be reinforced with two short 2 x 4s laid beside each other and surrounding the pipe. This will leave an overhang that must not interfere with siding, eaves, or interior ceilings.

When 4-inch pipe is used, a 2 x 4 wall (3½ inches deep) is obviously too small. So, an oversized interior wall is usually used for the pipe. It could be a common wall between kitchen and bathroom on the first floor and a bathroom wall upstairs. Of course, this is where it is important to plan rooms that will have plumbing, including the laundry, to be as close to each other as possible, or better, adjacent, so the common wall will hold most or all the plumbing. Such a wall has a 2 x 6 or 2 x 8 floor and top plates, either with studs the same size as the plates or a double row of 2 x 4 studs with their long side parallel to the plates so little cutting will be required.

Plumbing should be planned so that a minimum of cutting of floor and ceiling joists is required. In a cellar, the pipes can run below the joists, particularly if the ceiling is high enough for future expansion, or if the cellar is not planned for living space.

Joists should be cut only where the effect of their decreased strength is minor. When it is required, reinforcing plates of plywood running 2 feet on each side of the cut is a good way to assure full joist strength. Holes should be made only in the quarter end of the joist, and should not be more than 2 inches in diameter, and not less than 2½ inches from the top or bottom of the joist. If larger holes are required, the joist can be reinforced as above or a series of headers installed to compensate for loss of strength. At any rate, keep holes, and their sizes, in joists to a minimum.

About the only requirement for reinforcement of fixtures is the bathtub. Steel and fiber glass tubs compete against cast iron, but whatever the material, a tub full of water and containing a bather can be very heavy, and can cause sagging and all sorts of troubles. To prevent this, joists at the outer edge of the tub, or the end if the tub is at right angles to the joists, should be doubled. And when needed, the intermediate joist for the tub should be offset to allow the drainpipe to fit. The wall along which the tub sets should have extra studs nailed to the regular studs, cut short so they can support the lipped edge of the tub, and offset slightly (brought into the room a bit) so the lip of the tub can set on their tops. Regular studs also should have blocking just above the edge of the tub to act as nailers for dry wall or plaster lath.

Horizontal drainpipes must have a minimum slope of ¼ inch per foot. Too steep a slope will allow liquids to flow faster than solids, increasing the chance for clogging. The correct slope keeps the solids and liquids flowing at the same speeds, making the pipe self-scouring.

Finally, good plumbing provides vertical air chambers attached to water pipes in the wall at each fixture location, except the toilets. These chambers act as pressure relief valves to prevent sudden stoppage of water flow from hammering, which is not only annoying but can break connections and even burst pipes.

Heating

What kind of heat to use? Study this problem long and carefully, whether the climate is cold or warm. Heating systems can be just for heating, others can work for cooling as well. Also investigate the fuel to determine its price, availability, and ease of delivery.

The future in heating systems may be the sun — solar heat. The technology is new, it is expensive initially, but in certain climates it may be worth investigating. Conventional heating includes hot water, hot air, steam, and high-resistance electrical. Hot water, hot air, and steam can be fired by oil, gas, or electricity. No matter what the fuel, it seems likely that prices of all of it will continue to rise. The more insulation you put in your house, the more you'll save.

Concerning fuel, it is important to point out that while electricity is clean and efficient at the point where it heats, fuel must be burned (oil, gas, coal, or nuclear) to create the electricity, and such burning of fuel is highly inefficient. If you live in an area where electricity is created by water flow (hydroelectricity), then that might be the way to go. Otherwise, gas or oil may be the choice. Electricity generally is more expensive than other fuels.

Gas is usually piped into the house directly, and gas promoters point out that there are no delivery problems. When the only gas available is bottled, the problem of storage comes into consideration.

Oil must be delivered by truck, and is stored in a steel tank. Most oil tanks hold 275 gallons, and are placed in the basement, a disadvantage if you want to use the basement for living space. Larger tanks can be buried in the ground, accessible to the heating system and to the truck that delivers it.

Another thing to consider is how you heat: hot water, hot air. Steam systems are not as popular today as they once were.

Hot water: called by the hifalutin name of hydronics, this system uses a central water heating facility, and pipes it, usually by a pump (forced hot water) to baseboard radiators. It is an even heat, requiring pipes and baseboard units, installed by a heating plumber. A good point about this type of heating is that domestic hot water (for washing and bathing) can operate off this system, so the same fuel is used for heating and hot water. In nonheating times of the year, it is possible to heat the domestic water without heating the house.

Plumbing for a hot-water heating system sounds complicated, but it isn't, because the baseboards take up most of the space for piping above the floor, so there is not as much cutting of joists or exposed pipes in the ceilings (for second-floor radiators) as you might expect. Plumbing rules apply for hot-water heating pipes. Sometimes radiators are used instead of baseboards, but the same principle applies.

Hot-air systems are a fast heat, perhaps a bit less even than hot water, but efficient. The air is heated by a central furnace and forced by fans to ducts with outlets in the proper places, on exterior walls. Incidentally, efficient heating requires room units to be installed on outside walls, preferably under windows, to allow proper flow of air and prevent cold spots. Warm-air inlets should not be high in the wall, because warm air rises. When it cools, it returns to the main furnace by cold-air-return outlets and ducts, is filtered and recirculated. This type of system is basically closed, and reuses the air.

Gravity systems, in which hot water or air rises to heat rooms, without a boost by pump or fan, are obsolete because of their inefficiency.

Don't let anyone tell you that one type of heat is cleaner than another, or that one type of fuel is cleaner than another. It is not, at least concerning the interior of your house. Basically all systems discussed here simply heat the air, so if the air is clean, the heat is. Fumes and smoke go up the chimney into the air outside. Hot-air systems have filters that must be replaced regularly. One type of heat is not necessarily drier than another, either, although steam heat releases water vapor into the house. Hot-water heat uses sealed pipes, and no water or vapor is released.

All systems should have some kind of humidification if you find the house too dry during the heating season. The resulting dryness makes your nose and mouth too dry, can cause respiratory difficulties, and furniture and floorboards become loose when wood dries out and shrinks. Hot-air systems are most easily humidified by installing a humidifier at the source of heat, which circulates it through the house via the ducts. Other systems may need a separate central system of humidification, or the owner can use individual appliance humidifiers, one for each floor.

As for air conditioning, hot-air promoters are fond of suggesting that ducts can be used for cooling as well as heating. Yes and no. Hot air ducts can be used for cooling, but because they are, or should be near the floor for heating, their location is not good for cooling, because cool air falls. So, ideally, cooling ducts should be in the ceiling or in the wall near the ceiling.

Modern heating and cooling combination units use a fast-flowing stream of air to reach all parts of a room, and its promoters claim that the same ductwork, in the form of a fairly narrow flexible pipe, can be used for both heating and cooling.

A hot-air system uses an expanded plenum system. The plenum is the chamber where the air is initially heated, and when it is warm enough, a sensor turns on a fan, forcing the warm air to where it belongs. The expanded plenum is simply an oversized duct from which smaller ducts and risers branch out to different rooms.

Regular ducts are rectangular, and are designed to fit between studs in the wall, where they rise to the second floor. Other ducts are round. The framing must take ducts into account. Where the floor joists are perpendicular to the wall, the duct goes up through the floor near the wall and a baseboard unit diffuses the air into the room.

Incidentally, not too many years ago these baseboard diffusers were long, maybe 3 or 4 feet, even though the duct opening was not more than 10 inches. This was a ploy by the hot-air promoters to make the diffusers look like hot-water baseboard radiators, because hot water was the "elite" of heating systems at that time. Now diffusers are short, about 18 inches, and do a good, inconspicuous job.

Where a wall is parallel to the floor joists, the duct must go under the joists before it comes up through the floor. First-floor ducts do not have to come through the floor plate because modern diffusers provide for covering an opening in the floor next to the wall. But for second-floor heating, the risers must come up through the floor plate and top plate if they are to be concealed in the wall, and not interfere with the first-floor ceiling when they come out in a second-floor room. Upstairs diffusers can be the wall type or the floor type, with the air opening in the wall in both cases.

Whenever ductwork requires cutting of floor or top plates, a header should be placed between joists that are perpendicular to the wall and an extra joist installed between the joists flanking the duct. This will prevent sagging of the floor in front of the duct.

In the case of joists parallel to the wall, two blocks should flank the duct, connecting the stringer joist with the first inside joist, with a joist connecting them in front of the duct. Cold-air returns can be separate ducts, or can use the space between the joists where practical.

One important consideration is that you should never have unlined risers in a wall without insulation because that can cause considerable difficulties with condensation and exterior paint failure.

Ducts are made of galvanized steel and are put together with slip joints and sheet metal screws. A better duct system is made of firm, semirigid fiber glass, with a vinyl outer skin. While more expensive, they allow very little heat loss and reduce sound coming from the system.

chapter 18

The Overcoat, Thermal Undies, and Ear Plugs

Insulation –Against Cold, Heat, and Sound

Some Sunday afternoon, if you live in an area of snow, and the spirit moves you to take a walk or drive after a heavy snowstorm, notice the roofs of the houses around you.

Some will have a thick layer of snow on the roof. Others will have little or no snow, usually by the eaves.

Houses with snow on the roofs are well insulated; those without snow are insulated little or not at all. What happens is that the house heat escapes through an attic floor with no insulation and through the roof, melting the snow. Houses with insulation not only prevent heat from escaping, but can utilize the snow layer as added insulation.

Which all proves the value of insulation. And not just in cold climates. In hot climates insulation is growing more and more important because of the increase in air conditioning. It wasn't too many years ago that mild- or warm-climate houses didn't have any insulation because it was considered an unnecessary expense. For instance, in mild climate (warm winters, warm but not hot summers) like the Pacific Northwest, contractors poohpoohed insulation. But now, since even a little heating is very expensive, and air conditioning is also more expensive, maximum insulation with all its attendant benefits, is essential (Figure 45). The initial cost in proportion to the cost of the entire house is ridiculously small, with savings made up in a matter of a few years by reduced heating and cooling costs.

And today, with sound pollution a major factor in mental health, or lack of it, insulation acts against sound as well as heat and cold. Not only sound from the outside (traffic, airplanes, dogs, factories, etc.), but sound from household equipment and members of the family.

All building materials — that is, materials making up the actual structure of the house — ranging from wood, a good insulator, to aluminum, a poor insulator, insulate to some degree. Even air is an insulator, and the air space between a wall covered on both sides but without insulation in the cavity has insulating value. But because there are leaks of air into this cavity, and air movement simply due to the temperature difference between the outside of the wall and the inside, the value of air space is reduced.

So, insulation is introduced to trap air and hold it. One of the best materials is spun

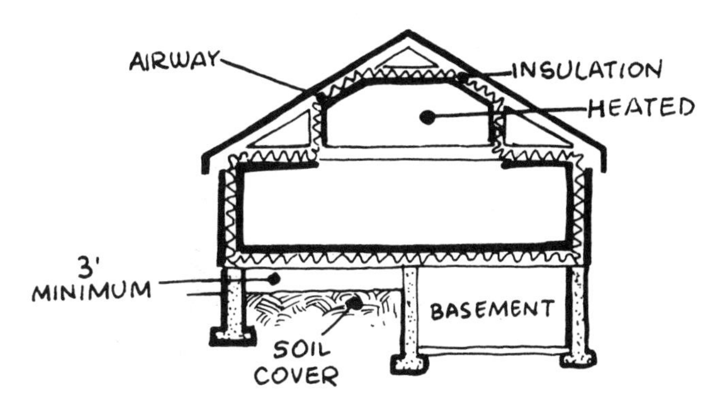

FIGURE 45. Insulation should go anywhere there can be heat leaks to the outside. Six inches in the attic floor, 3½ inches in the wall, 3½ inches over an unheated crawl space (allow 3 feet of space between the bottom of the floor joists and the soil, and include a soil cover of polyethylene or roll roofing). Soffit ventilation allows air movement from eaves to attic, where air escapes through gable vents. Always install a vapor barrier toward the heated part of the house. Recent recommendations call for 9 to 12 inches in attic floor.

fiber glass, though there are other materials, like spun or expanded minerals, vermiculite, urethane, and polystyrene.

Fiber glass is the most common and least expensive, taking into consideration its ease of installation. Urethane and polystyrene have inherent fire hazards, and the insulation boards, made of rigid materials, are quite expensive, so we will consider spun fiber glass in our discussion of insulation.

Some building codes allow flammable insulation, but require that it be covered on both sides by a building material. For instance, a flammable insulation should not be used on an attic floor, because there is nothing covering one side. However, it could be used in a wall, where it is covered by sheathing on the outside and plaster or plasterboard on the inside.

Because of the importance of keeping moisture from attics, wall cavities, and crawl spaces, vapor barriers, or materials impervious or with a high resistance to passage of water vapor, must be included in the installation of insulation.

Fiber glass insulation is made in thicknesses ranging from 2½ to 6 inches, in 4-foot-long batts or longer rolls, and in widths of 15 to 23 inches, designed to fit between joists and studs set on 16- and 24-inch centers, respectively. The batts and rolls come with or without a vapor barrier, usually aluminum foil or kraft paper. Fiber glass also comes in a loose fill type, suitable for pouring between floor joists of an attic. This stuff is usually factory-packaged scrap, and is not always available.

How much insulation? For maximum protection, 3½ inches in the walls (to fill the 3½-inch width of a 2 x 4 stud), 2½ or 3½ inches in ceilings of cellars or crawl spaces, and 6 inches in attic floors or roofs. Manufacturers have somehow gotten it into their heads to call various thicknesses "standard thick," "full thick," and something silly like "super thick," all of which makes no sense at all. The thicker the insulation, the more insulative value it has, represented in factors known as "R," and "K," and "U." The R factor signifies resistance to the escape of heat. The U value is a measure of heat transmission between air on the warm side and air on the cold side of the construction. The K factor represents heat loss. It is not necessary to get all involved or confused with these numbers. The number of inches of thickness is the most important thing to consider.

Insulating engineers are now figuring that 5½ inches of fiber glass insulation in the walls and 9 to 12 inches in the attic floor would save considerable fuel. The 5½ inches in the walls require 2 x 6 studs (see page 51).

However, the 9 to 12 inches in the attic floor would be no difficulty. Either 9 or 12 inches would extend above the tops of the joists in the attic, so it would be impossible to install a floor, even for storage. That, too, is a minor problem. When installing 9 to 12 inches in the attic floor, just add 3 inches to the 6 inches normally planned, or double it. Do not cover the joist tops, because they should be visible so you can walk on them when you go into the attic for any reason. If for some reason you have to walk around the attic for a project, you could lay down plywood as a temporary floor. This would mash down the insulation, but it would return to its original shape when the weight is removed, unless the weight is kept on it for an extended length of time.

Where to install? (Figure 46.) Starting in the walls, all exterior walls must be insulated. Batts or rolls are stapled between studs, with any vapor barrier toward the heated part of the house. All vapor barriers are installed toward the heat. The vapor barrier has a flange on both edges for easy stapling. A common method of installation is to staple the flange on the face (narrow edge) of the stud. A better method is to staple the flange on the side of the stud, indenting the insulation (and vapor barrier) ¾ inch, so that ¾ inch of air space is left between the vapor barrier and the interior wall (Figure 46). If the vapor barrier is aluminum foil, this is the only way its reflective quality will work at all.

In fact, some techniques of insulation are based on reflection on both sides: reflection of heat on the inside and cold on the outside in cold weather, and reflection of heat on the outside in hot weather. Reflecting insulation alone is inadequate.

Batts and rolls of insulation also come without a vapor barrier, and are often called friction fit. This insulation is stuffed between studs without fastening, and then 2- or 4-mil polyethylene sheeting is stapled directly to the studs as a vapor barrier (Figure 46).

The success of insulation depends on every nook and cranny being covered, and the same goes for a vapor barrier in the walls. So, narrow spaces should be insulated, too. You can cut the insulation to fit, and then staple the vapor barrier to cover it. The vapor barrier should cover stud edges and headers, because vapor will go through many materials, including wood.

When polyethylene is put on the walls, usually the entire wall is stapled thoroughly, then door and window openings are cut open. Where insulation meets the floor and top plates, any opening should be closed and the vapor barrier stapled.

Attic floors, or ceilings below flat or peaked roofs, should have 6 inches of insulation (Figure 46). Now, here's where there is not complete agreement on the vapor barrier. Some experts have found that a vapor barrier in the attic floor, particularly when the walls are covered with a vapor barrier, makes the house too tight, unable to breathe. The general result of this is too much moisture in the house. So, some insulation people recommend it be omitted in the attic floor. With 6 inches of insulation, however, there is actually less penetration of water vapor into the attic than if there is none, because with no insulation there is nothing to stop the vapor. And usually spaces under roofs are ventilated, so any vapor that penetrates is dissipated through vents to the outside. In fact, spaces under a roof should be ventilated, so take care not to extend the insulation into the eaves (Figure 46).

Loose-fill insulation, at least 6 inches, can be poured between the floor joists of an attic. Don't try to pack the insulation down.

Where living space is built close under the roof, 6 inches of insulation goes between the

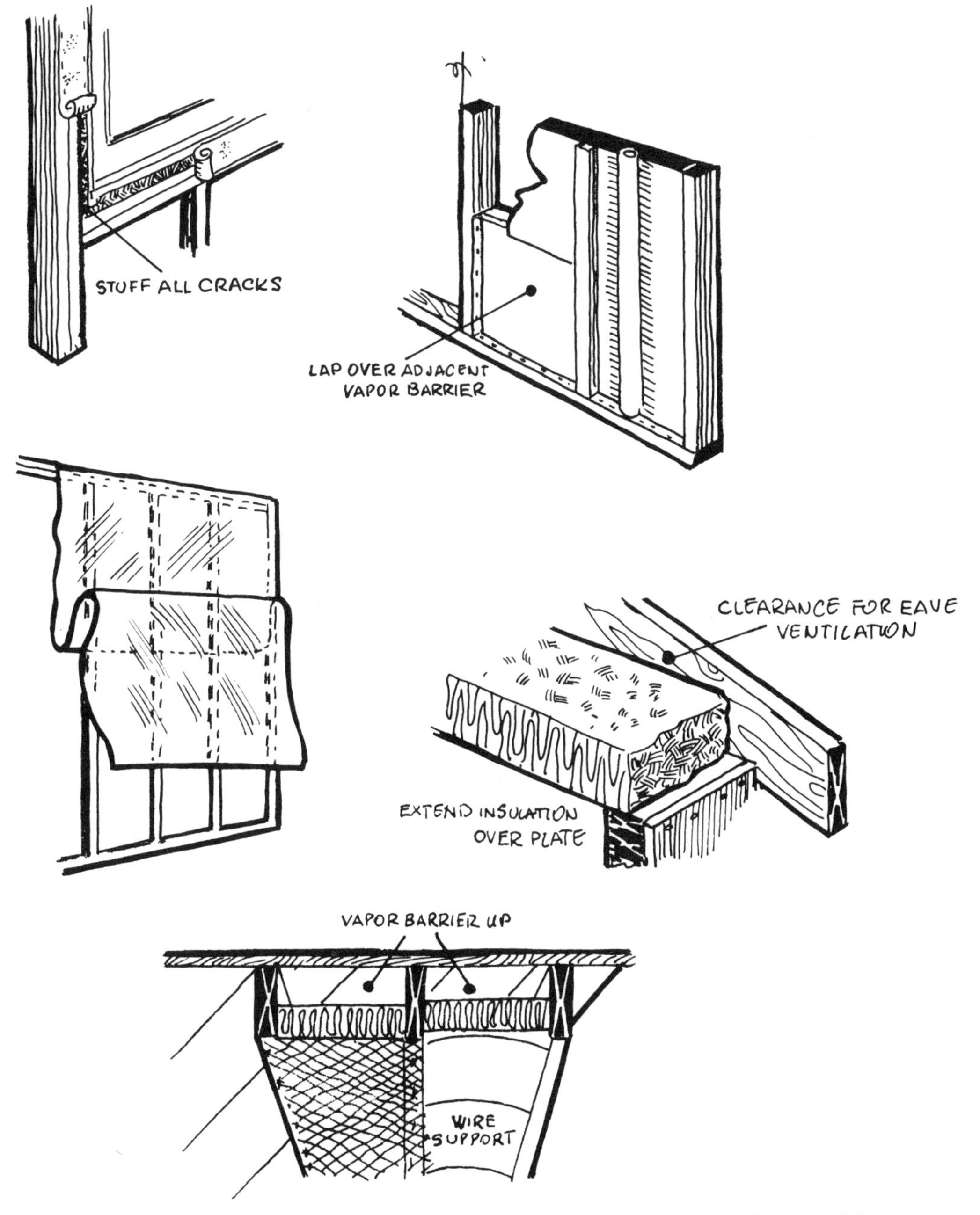

FIGURE 46. Insulating techniques. Fill all cracks with insulation and cover with vapor barrier (polyethylene, aluminum foil, or Kraft paper); overlap vapor barriers when piecing together insulation; put insulation behind pipes and electrical outlet boxes; cover entire wall with polyethylene when using unbacked insulation; extend attic floor insulation over the wall plate but not into the eave; leave a clearance for air movement if the soffit is ventilated; secure basement ceiling insulation with chicken wire or bent wire supports.

roof rafters, but at least 1½ inches of ventilated air space must be left between the insulation and roof sheathing. Here, the vapor barrier can be used, but still toward the heated part of the house.

In houses of more than one story, the stringer joists (perimeter of the house) must be insulated with 3½ inches of insulation and a vapor barrier. In crawl spaces, insulation is a must between floor joists (Figure 46). It also can be installed in cellar ceilings. For either place, it should be 2½ to 3½ inches thick. The thinner insulation is possible because cold air falls, while warm air rises.

Stapling insulation onto the bottom side of a subfloor can be difficult. The insulation must be pushed aside and the vapor barrier stapled. Because the insulation might fall off the vapor barrier by its own weight, staple chicken wire between the joists to hold it. Some insulation is made with a paper backing, complete with flange, but not a vapor barrier (the vapor barrier is on the other side), to ease installation, but some contractors don't think much of this because they think the paper could deteriorate, while galvanized chicken wire will not.

Vapor barriers are important to consider, not only in exterior walls, but also in crawl spaces. Dampness in crawl spaces can come up through the earth, causing structural members to mildew and rot, so the earth should be covered with a vapor barrier. Roll roofing, lapped several inches, or roofing felt, half-lapped, or 6-mil polyethylene is recommended. Try to install it so that it goes up several inches on the inside of the wall foundation.

Another method of installing a vapor barrier in, a wall or ceiling is to install plasterboard (either finish or as a base for plaster) with aluminum foil laminated to the side facing the outside of the house. Or the finish wall can be painted with one or two coats of aluminum paint, or a special vapor-resistant paint. In either case, such walls can also be painted with flat wall paint or wallpapered.

A vapor barrier must be toward the heated part of the house. Do not put an extra vapor barrier under the sheathing because you can get condensation between the two. So you must make sure that any sheathing paper or roofing felt is not impervious to the passage of vapor.

The vapor barrier keeps water vapor from inside the house penetrating the outside walls and causing exterior paint to blister and fall. Latex exterior paint is designed to breathe, allowing vapor to pass through it without blistering. This works only when the proper primer is used. And often, when latex is put on over oil paint, it won't stop vapor from coming through (if there is no vapor barrier) and blistering the oil paint and also the latex paint on top of the oil.

Slab floors, such as concrete slabs directly on the earth, should be insulated (Figure 12). If the entire surface is not insulated, at least the perimeter should be. This is usually done with rigid insulation board, made of fiber glass, mineral, or plastic.

In houses with so-called cathedral ceilings, which follow the roof line, and with exposed structural rafters, the ceilings must be insulated either between the beams or rafters or on top of the roof sheathing.

The effect of the beams is spoiled if 6 inches of insulation is installed between them, so rigid insulation (thicknesses of ½ to 2 inches are available) can be installed on top of the roof sheathing and roof shingles installed over that, using extra long roofing nails.

Sometimes, particularly with electric heat, where there are thermostats in every room for individual choice, interior walls and ceilings can be insulated. This to some people will make the rooms "dead still," perhaps too still, but it serves two purposes: to keep warm air from escaping a warm room through walls and ceilings into a cooler room, and also to insulate against transmission of sound. A vapor barrier in interior walls and ceilings is less important than in exterior walls and ceilings.

Ventilation goes hand in hand with insulation (Figure 45).

In an unoccupied attic, with insulation on the floor, lack of ventilation can cause overheated air in hot weather, reducing the effectiveness of the insulation. In cold weather, moist air can condense on rafters and cause rotting, not to mention dripping of water on the insulation, wetting it, and reducing or destroying its insulating value.

Another hazard in unventilated attics is a condition that can cause ice dams (see pages 86-87).

So, ventilators should be installed in the gable end walls, as close to the roof ridge as possible, to promote movement of air and keep moisture and heat from collecting in the attic. And at no time should these vents be covered, winter or summer.

The ratio of ventilating area to attic floor area is 1 to 300; that is, one square foot of total ventilating area to each 300 square feet of attic floor area. Louvers for protection against weather and screens to protect against insects and rodents must be subtracted from the gross ventilating area. Aluminum- and wood-louvered vents are available, and the bigger the better. You can usually determine the net ventilating areas of these vents. When additional ventilation is provided, as under the eaves (soffit of the cornice), the ratio can be reduced; that is, the area of ventilation drops to less than 1 to 300.

Eave vents can be installed in the soffit, and the work should be done during construction. A continuous, screened ventilation strip ¾ to 2 inches wide is ideal. Also, several long, narrow, louvered vents can be installed in the finished soffit. They will allow air to enter the attic through the eaves vents and exit through the gable vents. Vents also can be installed in the rake overhang, if there is enough overhang.

Hip roofs, which slope down to all four walls leaving no gables, require eaves vents, continuous if possible, and along all four sides. In addition, a roof vent should be installed, to allow air entering the eaves vents to dissipate through the roof vent.

The area of eave ventilation is not as critical as with gable vents, particularly if a roof vent is used. And if your house is 30 feet square, with a continuous 2-inch strip along the eave, you'll get 5 square feet of ventilation area on each eave. If your house is gabled, the two eaves will allow 10 square feet of ventilation. If your house has a hip roof, that's 20 square feet, plenty, not even counting gable or roof vents.

On flat roofs, where the roof and ceiling joists are combined, there must be free ventilating space between insulation and roof sheathing; a minimum of 1½ inches. And because the circulating air is restricted to the space between the joists, do not use solid bridging. For this same reason, the vents have to be only at the eaves that are perpendicular to the joists.

If the roof and ceiling joists are separate, the eaves ventilation can be similar. Roof vents add to the air circulation when roof and ceiling joists are separate. Where a flat roof has no overhang, ventilation can be in the side wall.

Another type of ventilation for a roof with gable ends is a ridge ventilator. The entire ridge is open for ventilation, with proper overhang for protection against the weather. This works best in conjunction with eaves vents.

Crawl-space ventilation is just as necessary as attic ventilation. We have already discussed soil covers (page 129), and repeat the need for them, as well as a minimum of 3 feet between soil and floor joists.

With a soil cover, the ventilation requirement is not great: 1 square foot of ventilation area for each 1600 square feet of crawl-space area. But when no soil cover is used, this ratio zooms to 1 to 160. Ventilators should be placed on all four sides (or three if the area is a wing to the house), near the corners. Cross-ventilation is a must.

In a basement, operable windows should be opened in summer (with screens). Ventilation is less necessary in winter. Where a crawl space is combined with a basement that has operable windows, vents in the crawl space are still a good idea. Ventilation is important whether the crawl space and/or basement ceiling is insulated, except in winter, when vents can be closed.

Insulation against cold and heat has been around for quite a while. Insulation against noise is relatively new, and is designed to protect our aching heads against noise pollution both inside and outside the house.

Sound is simply vibration that can be heard. When sound waves strike an outside surface, they cause that surface to vibrate, which in turn vibrates the studs, which vibrate the interior wall and wham!, the vibrations hit you and you're annoyed.

Exterior walls, with sheathing, sheathing paper, and siding, take up quite a bit of this vibration. Fiber-glass insulation takes up more, and by the time the vibration hits the studs and the interior wall, it's pretty well reduced. Double-glazed windows also can help reduce sound, and the use of storm windows in conjunction with insulating glass additionally helps reduce sound. With single glass, storm windows are a must and are recommended in cold climates even with double glass, not only for sound reduction but against cold, or more correctly, against loss of heat. While aluminum is one of the worst insulators you can buy, this disadvantage is outweighed by the convenience and relative lack of maintenance of storms of this material. Besides, storm windows insulate by creating an air space between inside and outside windows.

With exterior walls fairly well protected

against noise, consider interior walls (Figure 47). Engineers rate the resistance of a wall to airborne sound by a number called Sound Transmission Class (STC). The higher the number, the better the resistance to sound transmission. For instance, a wall with an STC rating of 25 will allow normal speech to be understood quite easily. But a wall with an STC of 50 will stop loud speech.

plasterboard, the rating is hiked to 46, good. An addition of 3½ inches of insulation will increase the rating to an even more satisfactory level.

A wall construction with resilient channels will isolate the 2 x 4 studs, preventing the wall from transmitting the sound to the studs and then to the other wall. These metal channels are made in the shape of an S, and

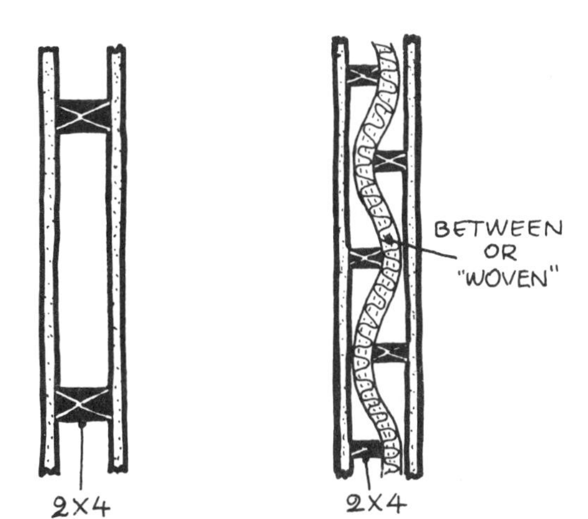

FIGURE 47. An interior wall with one layer of plasterboard on each side of studs makes a poor sound barrier; better is a double stud wall with 2 x 4 studs staggered on 2 x 6 floor and top plates with 2½ inches of fiber-glass insulation woven around studs. An exterior wall with thermal insulation is a good sound barrier.

Masonry walls are great sound barriers but in wood construction, they are impossible, or impractical. So, engineers have come up with various types of insulation for exterior and interior walls. Ideally, a wall should have a minimum STC rating of 45.

A 2 x 4 stud wall with ½- or ⅝-inch plasterboard on each side has a rating of 32 to 37, not enough. Adding an extra layer of plasterboard on both sides doesn't help much. But if you install sound-deadening board on each side, and cover it with ½-inch

should be screwed on both sides of the studs, horizontally, spaced 24 inches. Then, plasterboard is screwed directly to the channels with dry-wall screws. This can get kind of complicated, but the screw system of attaching plasterboard is very good, eliminating the popping of nails and other problems. A special drill must be used, but can be rented. Placing 2½ to 3½ inches of insulation between the studs will also improve the STC rating.

Of course, resilient channels on the out-

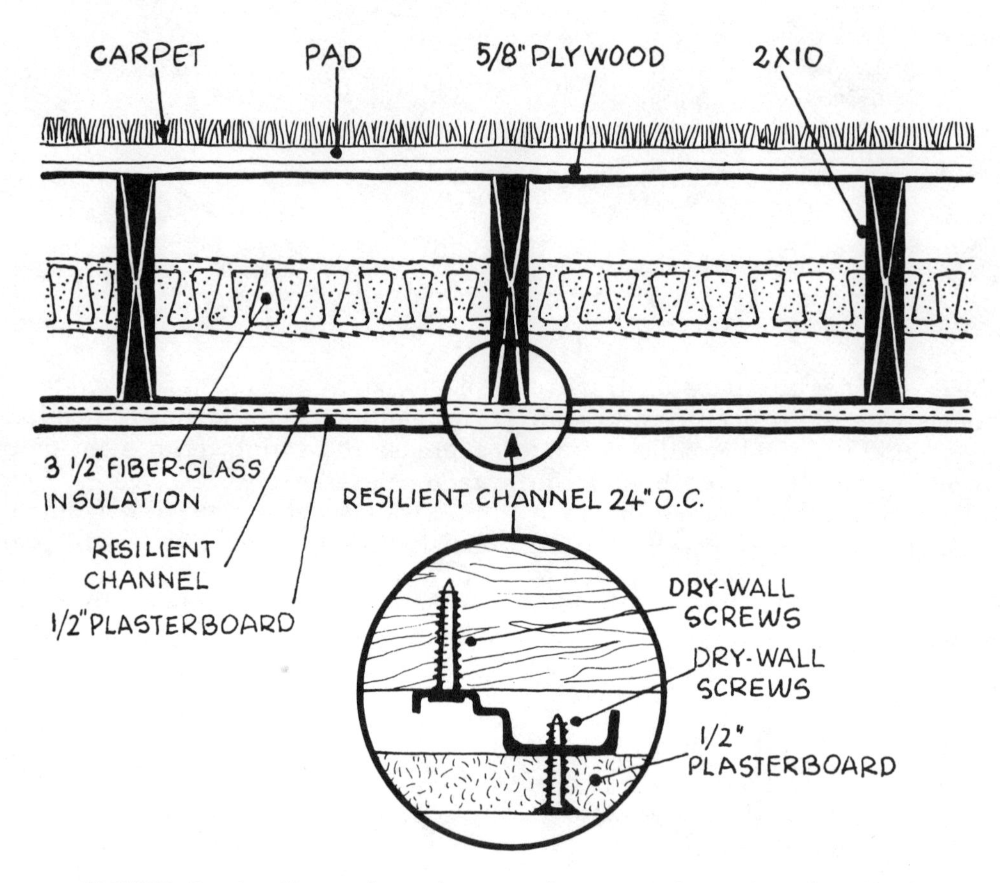

FIGURE 48. A ceiling with insulation in the cavity, plus a plasterboard ceiling attached with resilient channels (inset) makes a good sound barrier. Carpet and padding protect against impact noise.

side of an exterior wall may not be practical. And the sheathing and siding will help reduce sound transmission, particularly if the inside of the wall is fitted with the resilient channels.

A double-stud wall (Figure 47) is another way to reduce sound transmission and, while it is expensive, it might be usable in one area of the house, say where a bedroom is adjacent to the living room or any room is adjacent to the family room, from which most noise might emanate.

A double-stud wall uses 2 x 6 floor and top plates, and 2 x 4 studs set on 16-inch centers, staggered, so there is a separate stud system for each wall covering. When this system is used with ½-inch plasterboard, the STC rating is 45, very good. With addition of 1½ inches of insulation woven around the studs, the rating is increased to 49. Sound-deadening boards nailed to the studs and plasterboard glued to that brings it up to 50, even without insulation.

Ceilings present other problems (Figure 48). Not only do you have to consider the STC rating, but also the sound of impact noises, such as a child running around on a bare floor with his Buster Browns. Even vibration set up by machinery will be heard as impact — expressed by engineers as Impact Noise Ratings (INR) or, a more current term, Impact Insulating Class (IIC). INR is expressed in minus numbers (poor protection) and positive numbers (good protection).

A ceiling of 2 x 10 joists with a wood subfloor and wood finish floor, plus a ceiling of ⅜-inch plasterboard, has a fair STC rating but a lousy impact rating. Things aren't very well improved until the ceiling is isolated from the joists by resilient channels, spaced on 24-inch centers across the joists (Figure 48).

The most economical ceilings that are effective against sound and impact transmission are built this way: 2 x 10 joists, resilient channels with ½-inch plasterboard ceiling, 3½ inches of insulation between the joists, a wood subfloor, wood finish floor (or plywood underlayment) plus wall-to-wall carpeting with a foam pad. If you don't like wall-to-wall carpeting, fairly large area rugs with padding will also do well.

Of course, much of the transmission of sound through ceilings is from neighbors, as in an apartment house. A single-family house presents fewer ceiling problems, and a duplex (two-story apartments in a house side by side and sharing a common wall) presents only the problem of that common wall.

Another material used to reduce sound of interest to the single-family house builder is acoustical ceiling tile and panels, made from mineral or wood fiber and filled with many tiny holes or fissures to receive the sound vibrations and dissipate them by turning them into heat. Acoustical material is designed not to prevent transmission of sound, but rather to absorb it. Sound absorbed is not bounced back to you to smite your aching ears again.

If you plan to build a room with exposed beams, and there is a room above it, you may have to put up with more sound and impact transmission than you would like, particularly if the beams support the floor above. One sound dampening method uses 1-inch insulation installed just below the subfloor above, and suspended acoustical material below that, between the beams. Carpeting with padding should go on the floor above. If the beams are fake, or if they do not carry the floor above, then regular treatment of the ceiling-floor system is adequate.

Some tips on further reducing sound: electric outlet and switch boxes, as well as ceiling fixture boxes, should have insulation behind them. And they should be caulked with a resilient material, between box and wall covering. Boxes sharing the same wall, but on opposite surfaces, should not back up on each other, but should be offset 24 inches, or at least 12 inches. Medicine cabinets, if bathrooms back up against each other, should be offset so they are recessed in separate stud openings. If they are flush-mounted, it doesn't matter. Plumbing should be caulked where it penetrates the wall. The point of these precautions is to keep perforation of the walls to a minimum, to make sure the perforations do not coincide, and to seal all perforations as tightly as possible.

Another class of items to insulate are heating pipes, hot-water pipes, and heating ducts.

Heating and hot-water pipes can be insulated with material that comes in a tube cut in half, so the halves can be secured to cover the pipe, or with a wrap type material. The purpose of this insulation is to keep the heat in the pipes from escaping into the unused basement, particularly if the ceiling is insulated. Some contractors do not suggest insulating the basement ceiling, pipes, or ducts, pointing out that a basement warmed by such pipes or ducts will keep floors warm. It is our opinion that heat can be saved by insulating basement ceilings, and heat and hot-water pipes and hot-air ducts, and that a basement, even with its ceiling insulated, is not going to get intolerably cold in the dead of winter. The insulation will help keep the floors warm.

Ducts can be insulated against both heat loss and sound transmission. If you opt for hot-air heat, you can specify fiber-glass ducts, which insulate against heat loss and sound transmission. They are more expensive than metal ducts, but worth it.

Steel ducts can be insulated with duct insulation, 1-inch fiber glass with a vinyl backing, which goes on the outside of the ducts. The insulation comes in 4-foot-wide rolls, and is cut several inches longer than the circumference of the duct. It is wrapped around the duct, each edge folded, and the folds stapled with a special staple that locks into the two folded surfaces, holding them together. You can borrow or rent a gun that uses these staples from the place where you buy the insulation, usually hot-air heating outfits.

Cold-air returns do not have to be insulated unless the space they pass through goes below 40 degrees in winter.

Insulated ducts are more necessary with cooling systems than with heating. And with cooling, the seams must be sealed with duct tape.

chapter 19

Keeping Fires in Their Places

Chimneys and Fireplaces

Unless you use electric heat, a chimney is a must in a house. And even with electric heat, many house builders will opt for a fireplace, complete with chimney (Figure 49).

A fireplace is probably the most inefficient type of heat to use, but what the hell, it's romantic and fun, even though you get only 10 percent of the fire's heat into the room, and the other 90 percent goes up the chimney.

To start with, a chimney must sit on its own foundation and not be supported by any other structural member of the house. (See Chapter 7.) The top of the chimney must extend at least 2 feet above the peak of a ridged roof, at least 2½ feet above the peak if it doesn't penetrate the ridge itself, and at least 4 feet above a flat roof. Watch out for other roofs and trees nearby, for they can interfere with the chimney's draft, that is, its ability to funnel furnace and fireplace smoke and fumes into the air.

Most community building codes dictate the requirements for a chimney, but the general rule is that it must have a flue (an interior funnel in addition to the outside masonry), and it must be at least 2 inches away from framing lumber and ¾ inch away from finish material. The space between framing and finish must be filled with fireplace insulation.

An interior chimney — one that goes through the center or the inside of the house and penetrates the roof — is more efficient and cheaper to build than an exterior one. And its warmth helps warm the rooms it passes through.

Where it penetrates or extends beyond the roof, proper flashing must be installed, described in Chapter 14. The top of the chimney should be finished with concrete, and sloped, so that rainwater will flow off properly.

A chimney should have a flue liner of fired clay, which comes in various sizes and shapes. When using gas as a fuel, you must use a glazed clay flue, or stainless steel, because a lot of water vapor is created by gas burning, and when it condenses it can create diluted acids, which eat into other types of flue liners. You must determine the kind and size of heating equipment you plan to use in order to install the right kind of flue liner. Also, specifications are required on smoke pipes connecting furnace with chimney.

When more than one source of heat is installed, such as a fireplace, each heating

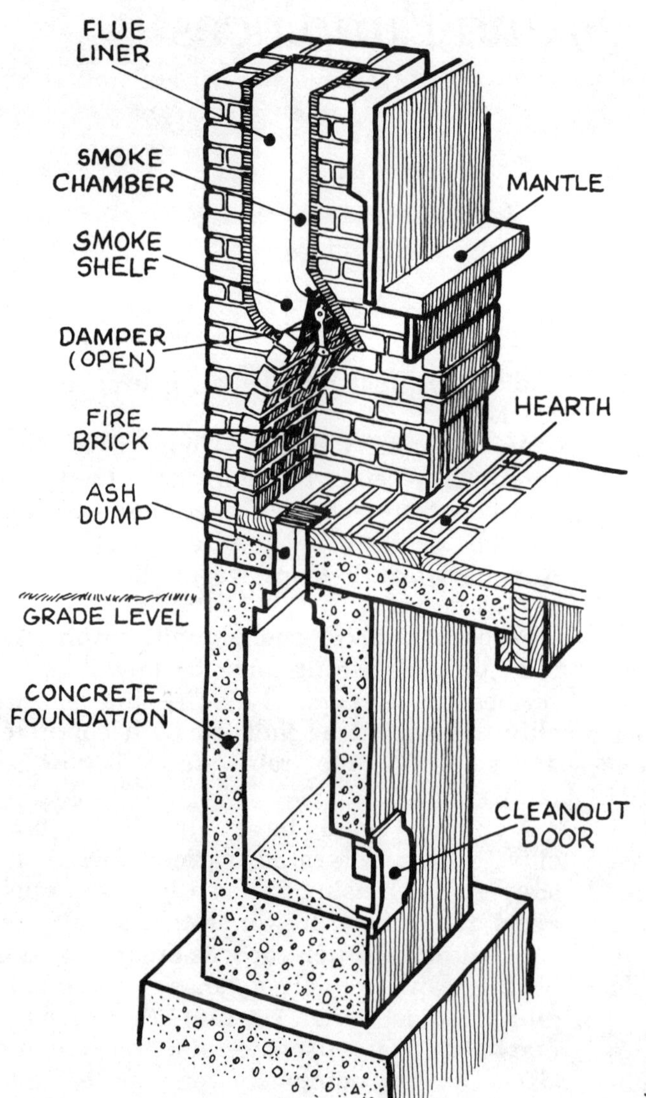

FIGURE 49. A fireplace, with a foundation, ash dump, and cleanout, properly proportioned firebox with damper and smoke shelf, plus flue liner in chimney.

source must have its own flue. One width of brick must separate each flue. Flue linings must be built ahead of the chimney (above the progress of the brick) so that the space between the flue liner and brick can be filled with masonry. Joints in flue linings must be bedded in mortar or fire clay.

Only fireplace flues can change direction. Furnace flues must be kept straight. Changes in direction must be made carefully, by cutting the flue liner or using specially shaped units. Above all, inside mortar of fire clay joints must be smooth to prevent soot and other combustion materials from catching.

There are quite a few things you can add to your chimney, such as a stone or concrete cap, to prevent rain and snow from coming in; a wire screen spark arrester that also keeps rodents out during nonheating seasons; rotating or conical tops to aim smoke in the direction away from the wind or to increase draft; and other goodies that make chimneys weatherproof, more efficient, and more attractive.

You can buy lightweight chimneys that do not need their own foundations, as well as prefabricated fireplaces, but make sure they have the approval of the Underwriters' Laboratories, Inc.

Chimneys must be kept clean, which is the best way to prevent chimney fires. Chimneys get dirty if you burn soft coal and a lot of soft woods, such as pine and fir, which create soot-producing pitch. Hardwood, hard coal, and gas and oil create relatively little soot. A chimney should always be inspected by a qualified chimney man, and cleaned, if necessary, by modern equipment. Chemicals advertised to clean out chimneys by throwing them on a hot fire are of little use, and can cause a chimney fire. If a chimney fire occurs, close all dampers and ash doors. to keep air from it.

Now consider that luxury, a fireplace. A very old heating system, it was replaced or improved on by old Ben Franklin, that Yankee genius whose stove brought the heat out into the room instead of letting it go up the chimney.

Then came central heating, after a few tries at various types of stoves, which relegated the fireplace to near-limbo. Of course, early fireplaces also were for cooking, and some of the "walk-in" fireplaces are quite common in very old houses.

If the fireplace wasn't eliminated altogether, it was a stroke of luck, because the Victorians, with their passion for efficiency, or something, blocked up a lot of fireplaces and even ripped some out, in favor of the monstrous fire in the cellar. But enough remained to restir the romance in people's hearts, so we still know how to make them.

There's an old feeling among fireplace builders that no matter how you follow the rules of dimension of firebox and flue, you still won't know if a fireplace will work until you build a fire in it.

Anyway, here are those rules: the depth of the fireplace should be about two-thirds the height of the opening. The flue area should be at least one-tenth of the open area of the fireplace if the chimney is less than 15 feet high; and one-eighth if the chimney is more than 15 feet high.

So, a small fireplace opening (it should be able to accommodate 2-foot logs no matter what its size), 20 to 30 inches wide, 28 to 30 inches high, and 16 inches deep, should have a flue of about 8½ x 13 inches or 10 inches in diameter if it's round; medium, 34 to 36 inches wide, 30 inches high, and 16 to 18 inches deep, and large, 40 inches wide, 30 inches high and 18 inches deep, a flue about 13 x 13 inches or 12 inches in diameter if it's round. As you can see, the ratio does not have to be exact, but should be fairly close if the fire is to draw properly.

Different designs of fireplaces must conform to the rules. If you opt for a fireplace open on two sides, or open on all four sides, the total opening must be calculated to determine the flue size.

You can buy prefabricated steel fireboxes that include the most important parts of a fireplace and, with warm air circulating around it, increase its efficiency. Even fans can be installed for further movement of warm air into the room. This idea should be considered, particularly if you like the idea of burning wood, to save on more expensive fuel. The prefab firebox will allow you to build bricks around it more easily, because the dimensions are already built in.

There are other things you must consider when building a fireplace. The front hearth is the floor of the fireplace in front of the firebox, and is built of tile or brick, and it must have its own concrete floor. The back hearth, where the fire is laid, usually has an ash pit with steel door. The fire chamber, or actual walls of the firebox, must be made of firebrick laid up with fireclay, a special type of mortar that is resistant to heat. Joints of fireclay are fairly narrow, about ¼ inch or less. The fire chamber sometimes has angled sides, with the back wall going straight up for part of its height and then sloping toward the room for the rest of the distance, to guide smoke and fumes through its throat and to help reflect heat into the room.

Where the firebox starts to taper to form the smoke chamber is a damper, which can be closed when the fire is out to prevent warm house air from escaping. Just behind the damper is a smoke shelf, an actual shelf that prevents smoke from doing a loop-the-loop and reentering the firebox. The smoke chamber tapers up to the flue. The top of the fireplace opening has to be supported by a steel angle iron set into the sides of the opening.

Fireplaces are quite simple once you recognize their function, but it still is an art to build them. You might be game to try anything in the house, but what you might do with a fireplace is to hire yourself out to the masons you hire to build the fireplace and chimney. What you'll do is haul brick, mix mortar, and carry the hod, and learn a great deal about brick, mortar, and fireplaces.

Brick is of various sizes, but standard brick is about 2 inches thick, 8 inches long, and 3½ inches wide. It is laid up in mortar, a mixture of Portland cement, lime, and sand that holds like crazy but has to be installed carefully. In fact, a good mortar will cling to surfaces even when they're held upside down.

You can buy mortar cement, also called masonry cement, which is Portland cement with the lime already in it. Mix 1 part mortar cement to 2 or 3 parts fine sand and enough clean water to make it workable. If you want to make your own mortar, mix 1 part Portland cement, 1 part hydrated lime, and 4 to 6 parts sand. For good-sized jobs, do not buy the ready-mixed material called mortar mix, which already has all the ingredients except water; it is simply too expensive.

What kind of brick to use? If you went to your friendly, neighborhood brick store, you'd be amazed at the type of bricks available: common, tapestry, glazed, waterstruck; the list is long. But it is a matter of taste as to the style of brick, because size is standard. Common brick is the least expensive and can be used both outside and inside for a rustic effect. The finish bricks are more decorative, and also a water-resistant brick is better to use outside.

Mortar must be laid up in all the joints, and it's tricky until you get the hang of it. A ⅜-inch joint is required, and bricks must be laid level and plumb. String a line at the right height once you get started, and make free use of a 4-foot mason's level, which can test for level and plumb surfaces.

Neatness counts in laying brick. Keep mortar off the face of the brick as much as possible. The stiffer the mortar, the less you'll smear it on the face. If you keep working it, it will get watery. Once it starts to set, in 15 minutes or so, strike the joints by dragging a pointing tool, an S-shaped steel tool used to smooth mortar. You can also use a sturdy piece of dowel. What you're doing is smoothing off the joint, making it slightly concave for a neat, weatherproof joint. In a day or two, or even weeks or months later, treat the surface of the brick with a diluted solution of muriatic acid to remove any mortar that got on the face of the brick.

Then you can look forward to using the fireplace someday. If it smokes or doesn't seem to draw very well, try reducing the opening of the firebox a little by putting a temporary board across the top of the opening. Once you determine the right size of the opening, by observing the behavior of the fire, you can make a permanent hood or decorative piece of copper to replace the temporary board. Even adding a layer of bricks to the hearth in the firebox can reduce the opening.

Another trick in improving the fire's draft is to open slightly a window opposite the fire, to create more air. If your house is well insulated and sealed, this might be necessary. Even a kitchen exhaust fan can affect the draft. If all these tricks don't work, you might have to put an extension on the chimney, or cap it, or do something to the top to increase the draft, or at least to aim the smoke with the wind instead of against it.

If you install the furnace thermostat in the same room as the fireplace, the heat of the fireplace fire will shut the furnace off, so you should locate the thermostat accordingly.

There are a few small prices to pay for the romance and fun of a fireplace, but once that fire is going brightly, you won't mind them at all.

chapter 20

The Inside Story

Interior Walls and Ceilings

After insulation goes into the walls, interior walls and ceilings are designed to be smooth, capable of being painted or wallpapered, or are prefinished.

Years ago, the only finish for ceilings and walls was plaster, and maybe, in fancy houses, wood paneling. The old-fashioned type of plaster had sand and horsehair mixed with it and usually had a lime base. More modern plaster has eliminated the hair and added gypsum, a natural mineral well suited to strong plaster.

Plaster is one of the best, and also one of the most expensive, types of plain wall surface. Generally, it is applied to lath, either a metal mesh secured to the studs and joists or wood lath, strips of wood 1½ inches wide and ¼ inch thick, with deliberate spaces. A base or brown coat of rough plaster is applied, forcing it through the mesh or into the spaces of the wood, to create a key to hold it in place. Then a skim or finish coat is applied, very thin, for a smooth surface. However, most modern plaster is applied over a 2 x 4–foot gypsum (plaster) board panel nailed to the studs and joists.

Plaster is usually applied by professionals, who are getting hard to find. If you do it yourself, the most important thing to do is apply plaster grounds to the wall at the floor, window and door frames, and sometimes ceilings. Plaster grounds are strips of redwood (to resist moisture and rot) ¾ x ⅞ inches, with the ¾-inch dimension as a guide to the thickness of the plaster and lath. The plasterboard lath is then put on and because it is ⅜ inch thick, the rough coat should be ¼ inch thick and the finish coat ⅛ inch thick to bring the wall to ¾ inches. Plastering is one of the dying skills in the house-building industry today, because of the popular, inexpensive dry-wall method.

Dry wall is the application of sheets, 4 x 8 and longer, of plasterboard to the walls that act as a finish wall (Figure 50). It is very inexpensive, and can be put up by an amateur, although certain skills are still needed. It can be nailed, screwed, or glued.

Let's start with the ceilings. In order to apply a nailing surface along the edges of a ceiling, 1 x 2 furring strips are nailed at right angles to the joists, every 12 or 16 inches. Use 8d nails. The strips can be shimmed with shingles to make sure they are level.

Where the joists are parallel to the wall, there is no place for the furring strips to be nailed, so when the wall is erected, an extra

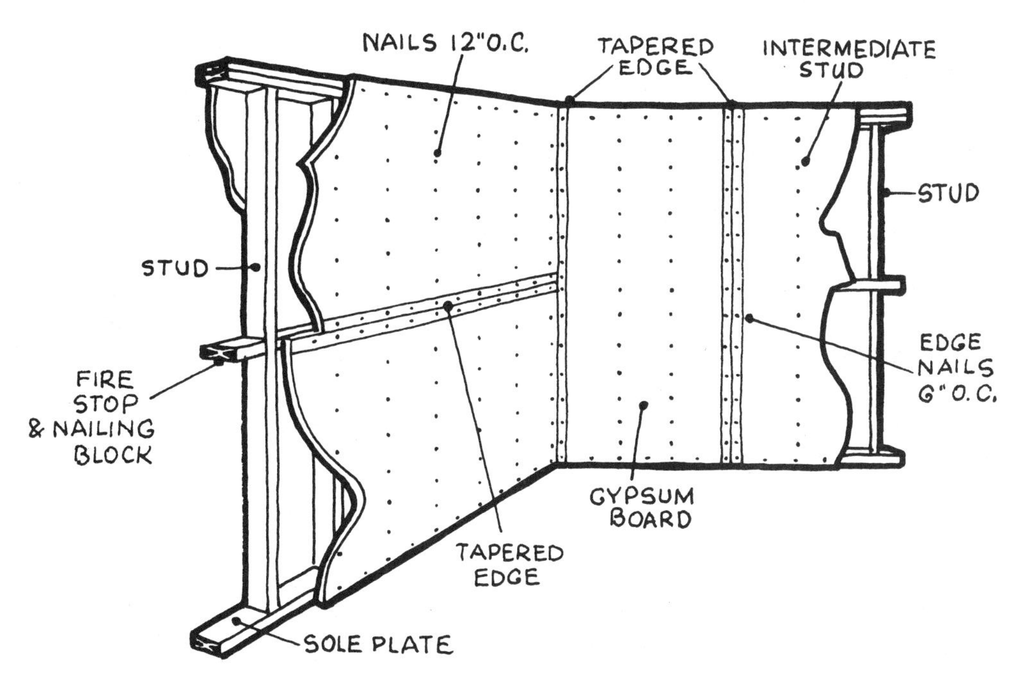

FIGURE 50. Two ways to install plasterboard: vertically, using 4 x 8-foot sheets, and horizontally, using 4 x 10-foot and longer sheets. In either case, nail spacing should be followed.

ceiling joist should be installed as a nailing surface. On a partition wall that is parallel to ceiling joists, short sleepers to match the depth of the joists should be installed to connect the joists on either side of the wall. Not only does this act as a nailer for the partition wall but the sleepers are there for attaching the furring strip at the wall. Another technique, when the wall is parallel to the joists, is to put a 2 x 6 plate flat over the doubled 2 x 4 plate. This 2 x 6 plate provides a 1-inch overhang on each side of the wall as a nailing surface. The same can be done on exterior walls, with the 2 x 6 overhanging to the inside of the top plate by 2 inches.

Standard thickness for plasterboard is ⅜ inch on 16-inch centered joists and studs, and ½ inch for joists and studs spaced on 24-inch centers. On the latter, ⅝ inch is better because it is stiffer.

Some amateurs will opt for two layers of ⅜-inch plasterboard, particularly on the walls, to give a good, solid wall surface, one that won't sound hollow when you tap it or bend when you lean against it. It also helps a little to reduce sound transmission.

Plasterboard comes in standard sheets 4 feet wide and 8 feet long. Sheets also come 6, 7, and 10 feet long, but the longer ones are a little harder to handle for one or two men.

For the ceiling, ⅜-inch plasterboard is the lightest to handle, but even that weighs 48 pounds for a 4 x 8-foot sheet. The longer the sheets, the fewer seams you'll have, which is important to the finish job.

To erect ceiling panels, a husky crew can help hold the sheet in place while it is nailed. You can also build a T-shaped brace a little less than ceiling height that goes under the sheet and holds it while you nail. But even with a brace, the erection of plasterboard is hard work.

Nail the sheets with plasterboard nails, which are blue-steel ring-shanked nails with large heads. Use the length required to secure the plasterboard well. Along the edges, nail every 6 inches, and nail flush with the surface.

On intermediate studs or strips, nail every 12 inches, and dimple the nailheads with an extra blow of the hammer. These dimples will be filled with joint compound.

To determine where the intermediate strips or joists are located, mark their position on the wall before you put up the sheet, and at each end of the sheet. Then, with the sheet nailed along the edges, you can connect these marks with a pencil or chalk line so there's no guessing where to nail. Too much guessing will create small holes where you missed and had to remove the nail, which can be a pain in the neck when you're trying to use joint compound.

Butt the long edges together. The edges are tapered slightly, so when they butt they create a valley. Then, when you apply joint compound to fill the joint, the compound can be smoothed off without creating a hump.

Where ends butt, or where you have to cut an edge and there is no taper, leave a gap at the joint of ⅛ inch. This way you can fill it with joint compound, and it will not only act as a key when the compound dries, but it also allows an adequate amount of compound to be applied with a minimum of hump.

For walls, the plasterboard can be ⅜ or ½ inch thick, with the ½-inch stuff preferable. And the sheets go up the same way. On walls, it's best to apply the longest sheets possible, horizontally (Figure 50), butting the lower sheet against the floor and cutting the upper half to fit under the ceiling. Ideally, the top half is put up first and the bottom half cut and fitted against the floor, but unless you have a crew to help, it's tough to get the top sheets up.

Laid up horizontally, the sheets provide just one horizontal seam. At doors and windows, cut out notches to avoid vertical seams at the top of doors and windows and at the bottom of windows. This is where most stress can apply when the house settles, and a solid piece of plasterboard will tend to crack less than a plastered seam.

To cut plasterboard, score the paper with a utility knife. The blade can be replaced cheaply. Don't use a saw. It will create a lot of unnecessary dust and the plaster will dull the blade.

Break the plasterboard at the score and cut the paper on the back side where the fold is made. It will break and cut cleanly.

Cuts also must be made for electric outlets, switch boxes, and ceiling outlets. The boxes must be nailed so that they extend away from the studs the same distance as the thickness of the plasterboard. So you must determine what thickness plasterboard you will use by the time you have the receptacle boxes installed. The box edges must be flush with the plasterboard surface so the box plates will fit properly.

To measure for a box, line the panel up with the box and use a square to extend the horizontal edges of the box to the plasterboard panel. For vertical edges, you must make two measurements: one from the point where the edge of the plasterboard will go to the near side of the box; the second measurement to the far side of the box. Transfer both measurements to the plasterboard panel. To make sure the corners of the box are square, make the measurements at the top and the bottom of the box, so you can make two marks to connect, both top and bottom.

Connect the marks and cut the box hole with a utility knife or hole saw. Cut it ⅛ inch larger on all four sides so you can maneuver the panel into position. The box plate will cover the enlarged opening. You must also cut out for the ears of the box, little studs at the top and bottom that are threaded to receive the bolts that hold the outlet or switch fixture.

Screws can be used to secure the panels. In fact, in many commercial applications they must be used because steel studs are used. The screws are self-tapping with Phillips heads, so a special driving tool must be used. One advantage to screws is that the work is faster and the screws can be placed on 12-inch centers.

Gluing is another good technique. Glues are getting stronger with the advancement of building techniques, and are conveniently packaged in cartridges to fit a caulking gun. A bead of glue is applied to the studs or joists or furring strips and the plasterboard pressed into place. Sometimes the panel has to be pulled off for a few seconds so the adhesive will set, and also to make sure all contacts are made. But once the glue sticks, it's a great holder. Sometimes, particularly on ceilings, nails are applied very sparingly, just to keep the panel in position while the glue starts to hold.

The advantage of gluing is that there are no nail heads to cover on intermediate nailing surfaces and no, or few, nails to pop.

Putting up the panels is a cinch compared to filling the seams and nailheads. This job takes practice, but an amateur can develop into a pretty good seam finisher (Figure 51).

Joint compound, which is a plasterlike material with glue added, does the job. WARNING: Many joint compounds contain asbestos fibers, which, when airborne and breathed, are extremely hazardous to your health. When sanding joint compound after it has dried, use a respirator approved by the U.S. Bureau of Mines, or instead of sanding dry, smooth with a damp sponge. The joint compound may have asbestos fibers even though the container does not say so. Someday, we hope, joint compound can be made without asbestos fibers.

Get the premixed joint compound, which is more expensive than the powder that you mix with water, but is of the right consistency. The compound is used plain over nailheads and with 2-inch paper tape over seams. The tape prevents the compound from cracking.

Apply compound with a smoothing knife. A 5-inch knife will do for applying the compound over nailheads; give it one swipe, smooth it off. For seams, apply it thickly, then smooth it off with a 10-inch knife. Apply the tape, pressing it right into the compound and removing excess compound. Finally, give it a skim coat of compound.

When you use the wide knife, work in long, steady strokes. Don't paint with the knife; you'll create unnecessary ridges. Use of the

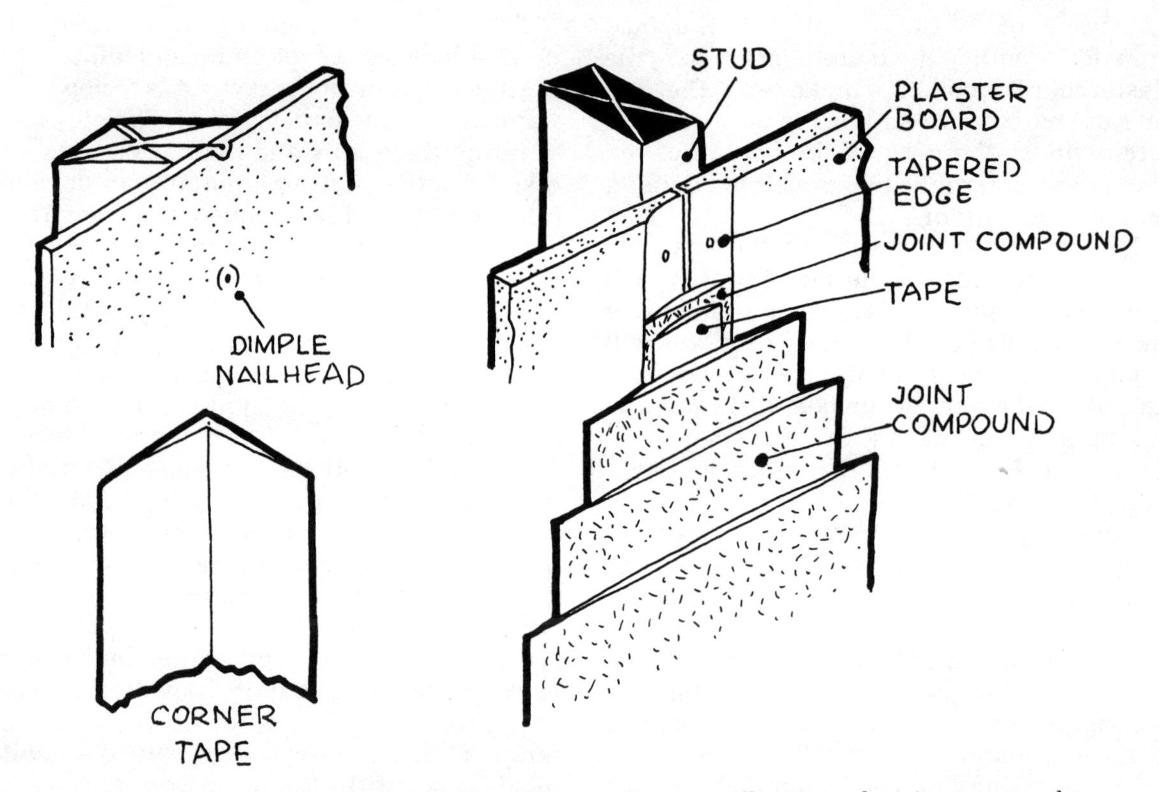

FIGURE 51. Dimpled nailheads, made with a hammer, allow space for joint compound to be installed. Tapered edges of plasterboard floor a valley to be filled with joint compound and tape, plus 3 skim coats of joint compound.

wide knife will almost automatically feather the edges of the seam; that is, make the compound thin in the center and taper it superthin at the edges.

Let the first coat dry. Apply a second coat to both nailheads and seams. Apply a third coat after the second dries. With each succeeding coat, taper the compound further out until the seam and compound is at least 10 inches wide.

If you get too thick a coat at the edges, remove about an inch-wide strip on each side, then smooth off again with the wide knife.

When all three coats are dry, the compound must be smoothed. Sand only with an approved respirator, and better yet, smooth with a damp sponge to remove ridges and feather the edges to an imperceptible rise. The better job you do of applying the compound the less smoothing you'll have to do.

In corners, vertically as well as where the wall meets the ceiling, the application is the same except the tape is folded lengthwise into an L shape (Figure 51) and pressed into the compound. When applying the compound over the tape, do so only on one side and let it dry. Otherwise you can gouge the side you did first. Corner knives, shaped like an L, are available, but some people might not like them or their price. While vertical corners should be taped and compounded, a molding or picture rail can be applied to the wall at the ceiling to cover the joint.

Where a closet or other construction sticks into a room, outside corners can be covered with a steel corner bead and nailed with plasterboard nails. The bead is resistant to bumps and is raised so compound is automatically tapered when it is applied to span the corner to the wall surface. Two coats are adequate here. Don't use paper tape. It is hard to keep straight and is not resistant to abuse. Steel corner beads are also used with regular plastering. You can also apply a wooden corner bead.

The walls are ready for painting or wallpapering.

Sometimes a ceiling is scrolled. This can be done with a skim coat of plaster or compound or with texture paint, a superthick paint that shows all the brushmarks. It can be applied with brush or roller, but with a brush you can make any kind of design. There are other types of walls and ceilings besides smooth ones. Ceilings can be tiled or suspended, with smooth material or with acoustical material.

Tiles come in 12-inch squares and 12 x 24–inch panels scored in the center to look like two tiles. They are applied to furring strips spaced on 12-inch centers. To apply the strips, measure the length of the room in the direction of the joists, find the center on each wall, and connect the two center marks with a pencil or chalk line. If each wall measures 12 feet, the center line will divide the ceiling into two 6-foot sections. Nail the first furring strip into the joists at the center line, and then nail strips on 12-inch centers on each side of the center strip until you reach the edge of the ceiling.

It is rare that a room will be exactly 12 feet. And that complicates the application of furring strips. Say it is 13 feet. The center line divides the ceiling into two 6½-foot sections. The procedure is the same: start a first strip along the center line and nail succeeding strips on 12-inch centers on each side of the center line. The second to last strip will be about 6 inches from the edge.

Another situation. Say the ceiling is 12 feet 4 inches: the halved sections would be 6 feet 2 inches each. A 2-inch border would not be so good, so to get around this, move the center line 6 inches to either side of the original. That way you divide the space into 6 feet 8 inches on one side and 5 feet 8 inches on the other. Thus the second to last strips will be 8 inches from the edge of the ceiling instead of 2 inches.

Now, with the furring strips up, you must determine how the tiles will go in the other direction. So do the same: measure the center of each wall and connect them with a pencil or chalk line. Make adjustments if necessary so that edge tiles will not be less than 6 inches wide.

Tiles are tongued and grooved, with the groove side having an extra flange for stapling. Because they are tongued and grooved, they must be installed from one side of the ceiling. Therefore, you must cut the edge and corner tiles so that they will coincide with the center of the strip next to the edge strip. And they must be nailed with finishing nails along the edge, because the tongues must be cut off so they'll butt against the ceiling. If molding is planned for the corner where ceiling meets wall, the tiles don't have to fit tightly.

Beveled tiles are designed so you don't have to butt the edges tightly. This gives you a chance to fudge a little, opening up the seams in order to keep the row you're working on straight. Other types of tiles are not beveled, and are designed to butt against each other tightly, to give an effect of a one-piece ceiling. You simply have to be more accurate in your planning and placing of the furring strips so you can butt the tiles tightly against each other.

A suspended ceiling is planned and measured the same way, except grid units instead of furring strips are installed to hold the panels in place. The grid must be at least 4 inches below the joists, and the ceiling is suspended, or hung. There is also a system of clips that can be attached directly to the joists, giving the ceiling a maximum height. A suspended ceiling is not good if the height from floor to bottom of joists is less than 7½ feet.

Some suspended ceilings are designed so that the grids are exposed and panels, usually 2 x 4 feet, are dropped into them, and can be removed for access to wiring or anything else inside the ceiling. Such exposed grid ceilings are not very appropriate for formal rooms.

Still other types of suspended ceilings take advantage of the grid system by making larger grid units and painting or covering them to look like wood, so you have sort of a beamed ceiling. Not very convincing.

A recent development is a suspended ceiling that uses grids, but instead of dropping into the grids loosely, panels are inserted in the grids and butted against each other for a ceiling that looks like one piece. Even with this system, the perimeter grid is exposed.

In both systems, the grids are hung from the joists with wires or clips.

Suppose there is a part of your house that you want done in wood paneling. There are still, if you're lucky, solid wood boards, usually pine, sometimes fir or cedar, that tend to be rustic. Then there are hardwood boards, very expensive but elegant. The advantage of these boards is that they make a very solid wall, with no backup material. Used boards, weathered for many years, are also available.

Less expensive and fairly good-looking are 4 x 8–foot panels of real hardwood laminated onto cheaper core lumber for a thickness of about ¼ inch. Finally there is the phony paneling made of hardboard, with a photographic covering of "wood grain," and the audacity of audacities, a phony photographic "wood grain" on real wood.

Paneling serves a good purpose, but be aware that it should be used sparingly. A whole room of paneling can be too much of a good thing, particularly in general living areas of the house. A real recreation room, often appropriately called a wreck room, may do well to have all wood paneling so it can take a beating. But in other places it should be used, say, as an accent wall, flanking a fireplace, or applied to a long, blank wall that needs something to relieve its monotony.

Solid board paneling is ¾ inch thick and comes in 1 x 6, 1 x 8, and 1 x 10 random widths. It's best with a beveled V groove. It is tongued and grooved for easy installation. It is usually put up over furring strips, 1 x 2 or 1 x 3 boards nailed horizontally every 24 inches on a stud wall. The paneling is nailed vertically into these furring strips. The boards are nailed with finishing nails, 8d or 10d, piercing the tongue so that the groove fits over the tongue to cover the nails. On extra-wide boards, 1 x 10 and wider, face-nail the middle of the board, countersink it, and fill with putty.

Board paneling is very sturdy. It will hold anything, such as shelves, heavy mirrors and pictures, anything you want to put on it. If it does need finishing, a clear finish like urethane varnish is good. Use semigloss; high gloss will glare. You can also stain it to your taste. Some stains are wood-toned, others are colored and include whites, greens, blues, reds, and yellows. The prefinished hardwood boards of oak, cherry, and walnut are so expensive that even an accent wall is very costly. But they are quite elegant.

For a different effect, you can apply boards horizontally. They can be nailed directly to the studs, so the furring strips are not necessary. Horizontal boards are effective, but those horizontal V grooves do collect dust.

A 4 x 8–foot plywood paneling is grooved to look like random paneling. The grooves are spaced so that no matter how random they look, a groove always lies on 16-inch centers, so that when you nail it, the nails will go into the grooves for an invisible fastening. The plywood comes in many species — birch, walnut, oak, mahogany, etc. — and is stained and varnished to a very smooth surface, light, medium, or dark in color.

The plywood and hardboard panels go up the same way. You can put them up on bare studs, or horizontal furring strips, the strips on 12-inch centers. (The panels are very thin, and have an unsatisfactory hollow sound when touched, bumped or rapped. And when leaned against, they yield in a most disconcerting way, giving a sense of cheapness and flimsiness. If you can tolerate this, fine.) You can nail the paneling with nails colored to match the color of the paneling. With horizontal furring strips, nail right into the groove. The nails are small-headed and ring-shanked for good holding power, and you don't have to countersink them. In fact, the panels are too thin for any kind of countersinking.

Nail every 6 inches, on edges, top, and bottom, and on intermediate strips or studs. Or you can glue the panels on with adhesive, which comes in cartridges dispensed from caulking guns. Follow instructions, which usually dictate that a bead of adhesive be applied to the furring strips or studs, the panel put into position and pulled off the adhesive to allow the adhesive to set for a few minutes, also to make sure a positive bond will occur at all critical contact points, and the panel reapplied, with nails top and bottom to hold it while the adhesive starts to do its work. The panel can be seated on the adhesive by tapping a hammer on a board held over the panel to protect its finish.

It is better to put paneling over a wall of plasterboard ½ or ⅜ inch thick for obvious reasons. And you don't worry about the nail heads or seams of the plasterboard, except to drive all nails flush. If necessary, you can put up underlying panels of plasterboard horizontally. Then, nail the paneling with extra-long nails, which will penetrate the plasterboard and go into studs or furring strips. You can also use adhesive; apply it in a zigzag pattern, then proceed according to adhesive instructions.

The point is that plywood paneling glued to plasterboard will give a good, solid feel, without the flimsy and hollow sound that unbacked paneling gives. Besides, gluing the paneling to the plasterboard helps with sound control.

In traditional houses, particularly Colonial, wainscoting sometimes is incorporated into the wall. Wainscoting means applying boards or other paneling, usually wood, to the wall, either all the way up to the ceiling or partway. It comes from the Old English, which in turn comes from Dutch words meaning, perhaps, timber for wagons. Its purpose was not only decorative, but to protect the lower part of the wall from the bumping of chairs and other furniture. Old-fashioned plaster was pretty tender, and could be damaged easily.

Wainscoting can be applied directly to the studs, with nominal 1-inch boards, vertical or horizontal, or with ¼-inch paneling, placed vertical and right onto the face of the plasterboard. Height of wainscoting is a matter of style. It can be as low as 2 feet or so and as high as 4 feet. When wood covering goes from floor to ceiling it becomes paneling. It can be topped with a chair-rail molding or any type of molding that suits your taste. A baseboard is not always used with wainscoting.

You can put up a chair rail without wainscoting, using a molding or an ordinary board (nominal 1 x 2 or 1 x 3) with chamfered (beveled) edges. The wall then acts as a sort of fake wainscoting. The chair rail can be painted the same color as the woodwork or left the same color as the walls.

In kitchens and bathrooms, ceramic tile

and laminated plastic are good surfaces for sink splash areas and shower and tub enclosures.

Ceramic tile is clay baked and glazed to a glass-hard finish that will last for more than the life of the house. You can also get real glass tile, equally long-lasting.

The old way of putting up ceramic tile was to nail up metal lath, lay a bed of Portland cement mortar, and imbed the tile in it. It made a permanent, waterproof tile wall or floor, but it wasn't the kind of job that an amateur would like to tackle. Today, with adhesives designed for the job, ceramic tile can be applied to plasterboard walls. The plasterboard should be water-resistant. Avoid putting tile on plywood, which expands and contracts according to its moisture content and tends to work off the tile.

Ceramic tile comes in many styles, from standard 4¼-inch squares to 1-inch mosaics applied to mesh in 1-foot squares. It also comes in oversized pieces of all shapes.

To install ceramic tile, measure the width of the wall and find its center. Make sure this center line is plumb. You can adjust the line so that border tiles are of maximum width. If the tub is not level, the first row of complete tiles must be applied far enough above the tub line so filler pieces can be cut to fill the uneven gap.

Once tile position is located, apply adhesive with a serrated trowel. Label instructions on the adhesive can will show how large the teeth in the trowel should be so that the proper amount of adhesive is applied. Then each tile, or sheet of tile, is applied. Individual tiles are applied with a slight twisting motion, but not enough to allow adhesive to ooze up between tiles. The 4¼ x 4¼-inch tiles have lugs on each side for automatic spacing. Sheet tile must be spaced the same as the space between the individual mosaic pieces. Every now and then, take one of the tiles off to make sure plenty of adhesive has been transferred to its back.

After all the tile is installed, a white or colored grout is applied. This is a powder you mix with water to the consistency of heavy cream and apply with a squeegee so that it fills all joints. Excessive grout is wiped off, and when it begins to set, the joints are struck by running a stick, such as a tongue depressor, down the joint, making a concave surface. You can use your fingers, but the grout is corrosive as well as abrasive, and is pretty hard on the skin. It is best to use rubber gloves when working with grout since the Portland cement in the product is dangerous and can be absorbed by the skin.

After striking, the film of dried grout is wiped off the face of the tile with a dry cloth, and the face of the tile polished with a dry cloth. Spraying with a silicone grout sealer will waterproof the grout and will help keep it the original color. At the tub line grout is also applied. Figure on renewing the grout at the tub line every three or four years. Instead of the cement-based grout, you can use a soft caulking material, but it gets very grimy.

To cut a tile, score it with a glass cutter and snap it, just as you would glass. Snapping of tile is tougher than glass, so set the tile on an edge such as a ruler and step on the part you plan to discard. Tile cutters can be rented or loaned when you buy the tile.

They are good, but sometimes they'll break the tile just where you don't want it to break.

Ceramic tile can also be used on splash boards behind a counter, behind stoves and ranges, and on countertops.

Another material for such uses, including tub and shower enclosures, is laminated plastic, such as Formica. It is put up in sheets, which come with molding, cut to go at seams and corners and along the tub edge, as well as on top of the material, unless it goes all the way to the ceiling. It lasts a long time, but its installation is trickier than that of individual tiles because the sheets must be cut to follow the contour of the walls, ceiling, and tub. Two pieces go along the side of the tub and one on each end. The plastic is inserted into slots in the molding, so the molding must go up at the same time as the sheets.

The sheets are put up with contact cement, a highly flammable material when wet, so use plenty of ventilation, turn off pilot lights, and allow no flames whatsoever. Contact cement is brushed or rolled on both surfaces to be cemented. It is allowed to dry, and the pieces are ready to put together when the cement loses its gloss and is dry to the touch. Once connected, it is virtually impossible to separate the pieces without wrecking the plastic or the material on which it is cemented. Caulk all joints to make a waterproof seal.

Plastic backsplashes go up the same way when kitchen cabinets are installed. If a backsplash does not end up under a cabinet, a cap molding is installed.

Any time ceramic tile ends in the middle of the wall, whether in bathroom or sink, horizontal or vertical, it can be capped with a special end piece of tile finished off to a tapered edge. Wood moldings also can be installed where ceramic ends.

chapter 21

Floor Show

Flooring

You've had a floor on — or in — your house since you set the floor joists for the first and second stories, and installed the rough flooring.

Now what you want to do is put on flooring, which, like siding, is the finished surface, designed not only to look good but to wear well, too.

There are many kinds of flooring, including wood, resilient (plastic tile), carpeting, and for special surfaces, ceramic tile.

Wood is still the best all-around flooring you can use (Figure 52). The most popular wood is hardwood, generally oak, but there is also birch and maple, and the more exotic beech and pecan. Unless you can go to a housewrecking outfit and get some old hardwood flooring, settle for oak, which comes in strips 25/32 inch thick and 2¼ inches wide (the widths can vary) or planks up to 5 or 6 inches wide.

The strips are tongued and grooved, and they are also end-matched, which means that their ends as well as their long edges are tongued and grooved. It also means you can place ends of strips between joists.

Oak flooring comes in various grades, from clear (no blemishes) to No. 1 common, which is full of knots and other blemishes, but which has more character. Strip flooring also comes in fir and hard pine, but these softwoods are less satisfactory than harder oak.

Before installing strip flooring, put down sheathing paper or roofing felt, overlapping just a few inches. Strip flooring must be installed at right angles to the floor joists, so that each nail will go into a joist. This is no trouble when the subfloor is plywood or boards nailed diagonally. But if the subfloor boards are at right angles to the joists, a second layer over them will tend to cup and not behave very well despite thorough nailing, because even thorough nailing will not hold well in a subfloor. You could use screws, which would help a little.

Because wood expands and contracts with the amount of moisture in the air, the wood strips should be brought into the room in which they are to be installed at least two weeks before they are installed, so their moisture content can equalize with the moisture in the room.

The first strip is laid ½ inch from the edge of the wall to allow for expansion and contraction; otherwise the expansion will cause strips to buckle. If you put down the strips in the winter when the heating system is work-

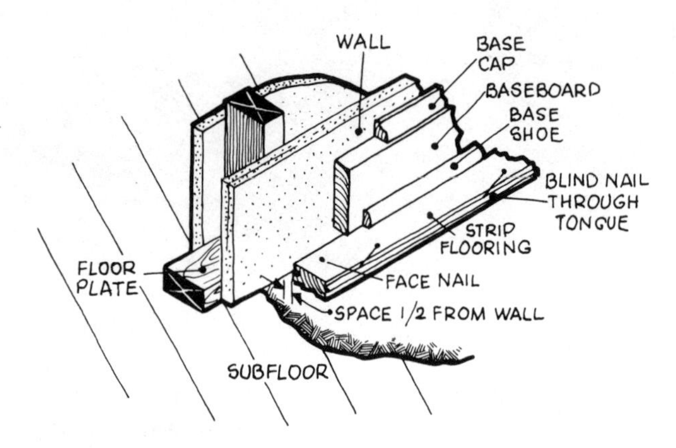

FIGURE 52. Start strip flooring ½ inch from the wall. Baseboard goes on top of the flooring, with a decorative cap and base shoe (quarter round).

ing and the moisture is at its minimum, the boards will be at their smallest dimension. Then, when they swell in the summer when moisture is higher, they will stay in place more tightly, if properly nailed.

Face-nail the first strip, driving the nail through the face of the board. You may have to predrill nail holes because oak tends to split. Use flooring nails, cut nails long enough to go through the subflooring and into the joist by at least an inch. Cut nails are made from steel and are square or rectangular, with blunt ends, which allow them to crush their way through the wood. Pointed nails tend to separate the wood fibers, causing the wood to split. The first strip is also nailed through the tongue, at an angle, with the nail set flush with the tongue. A special nailing tool can be used for speed and convenience, and will help prevent splitting.

The grooved side of the second strip is then forced onto the tongue, and the tongue of the second strip is nailed. The strips come in various lengths, and the longer the strips, the more expensive the flooring. Short strips, if they are end-matched, add to the character of the floor. Each strip is completed in place; that is, from wall to wall.

Sometimes a strip will be slightly bowed. Then you have to use brute force to snug it up against its neighbor. You can use a spare strip to act as a striking surface so you don't damage the tongue or surface of the strip you're installing. You can also cut scrap boards about the length of the room from the strip you're working on to the opposite wall, and force them into place, moving each strip snugly against its neighbor. Sometimes boards can be nailed temporarily to the floor, close to the strip being installed. Then use a prying bar to lever the strip into the position, using the temporary board as a prying surface.

The last strip to be nailed may have to be ripped (cut lengthwise) to fit. At any rate, its edge must be at least ½ inch from the wall, for expansion purposes. This ½-inch gap is less critical at the ends of the strips because they expand and contract less along their length than across their width. A small gap still doesn't hurt. The last strip is face-nailed. The quarter-round shoe mold at the bottom of the baseboard will cover any gaps.

Most oak-strip flooring comes unfinished, and you have to sand it to make each strip level with its neighbor, and to smooth the wood surface. Other strip flooring comes finished, even waxed, which is fine except the boards might end up with slight ridges at the joints.

To sand a floor, rent a floor-sanding machine and an edger. The sander is a monster machine, with a drum around which sandpaper is secured. You have to sand three times: with coarse, medium, and fine sandpaper. The most important thing to remember in sanding is that the machine should be moving at all times; otherwise that madly revolving drum could gouge or make the floor uneven. Work with the grain of the wood, parallel to the strips at all times.

The big sanders don't sand right up to the edge of the floor, so the edger is used. This is a small but powerful machine that uses a disk instead of a drum to do its work. Because the disk makes circular marks in the wood, the medium and fine sanding steps should be done particularly carefully.

Oak flooring also comes in planks, ranging up to 7 or so inches wide, and usually prefinished. It can be nailed, but because of the width must be face-nailed, at least near the groove side. Sometimes planks are screwed on, with a hole drilled for the screw shank and a larger one drilled for the head, so the screw can be countersunk and filled with a hardwood plug. Some plank flooring comes with phony plugs for decoration. These planks must be nailed.

If you go for Colonial style, you can put down wide pine boards, the wider the better. Any grade will do; either knotty or clear pine will be authentic.

Wide pine boards are usually not matched or tongued and grooved, so they must be face-nailed, with two nails in each joist. On extra-wide boards, three nails may be necessary.

Install these boards the same way you would the strips. Use 10d nails. They can be screw nails, which actually turn when you drive them with a hammer, or you can use long, ring-shanked nails. Galvanized box nails are also good to use. The latter have larger heads, which make them harder to countersink. And when they are countersunk, they make big holes. The holes can be filled with wood putty to match the wood or left as is; after all, wide pine boards are informal, at best. Despite this, they look very good properly finished and are particularly elegant with fine rugs.

It is doubly important to store pine for two weeks in the room in which it will be installed to allow its moisture content to adjust to that of the room. If you install pine boards in cold weather, when they are driest, they will stay together longer.

Wood floors, particularly oak, will last the life of the house. When they get pretty shopworn, after ten or fifteen years (thirty if you really take care of them), they can be refinished by sanding. Strips can be sanded four or five times.

A disadvantage of pine and other softwoods is that they tend to dent and gouge more readily than oak. But there are many houses, particularly in the East, that are two hundred years and older, and the wide pine boards are still in pretty good shape. If they are cupped or uneven, they can be sanded. Sometimes they can be taken up and reversed, then sanded. Be careful when sanding pine. It is so soft that a sanding machine can really damage it.

Sometimes you see pine boards in the floors or walls of very old Eastern houses that are 24 inches wide and even wider. These were most likely illegally come by when the United States was an English colony, for any tree from which you could get a 24-inch board was claimed as a king's tree, and was assigned to King George's navy for use as masts and spars in those magnificent frigates and ships of the line.

Back to the modern world. You can put wood floors over concrete, in the basement or on a slab on grade. It helps if the slab has been laid over a moisture barrier, which is 6-mil polyethylene laid right on the ground before concrete is poured (Figure 12). Then, 2 x 4 sleepers are installed on their wide side on the floor. If the concrete does not have a vapor barrier, apply the polyethylene to the top of the 2 x 4 sleepers, which are anchored to the floor with concrete nails or adhesive. They should be treated with a water repellent and wood preservative. It also wouldn't hurt to put a waterproof coating on the concrete: silicone waterproofing, roofing cement, or even hot tar. The concrete nails are hard steel, and are driven flush with the 2 x 4 surface. Don't try to countersink them, you'll end up breaking up the concrete. The sleepers can be shimmed with shingles to make them level and to make up for any unevenness of the concrete.

The sleepers are laid on 12-inch centers, and the strip flooring laid directly on them and nailed in the regular manner. If you put the sleepers on 16-inch centers, a plywood subfloor is required to prevent springiness.

Resilient tiles and wall-to-wall carpeting are two good flooring surfaces, but their life is limited. Tile will last ten to twenty years, carpeting perhaps eight to fifteen years. This is fine, except that if the cost of carpeting, particularly, is included in the mortgage, which might run for twenty to thirty years, you're paying for it long after it has worn out. Resilient tiles are not recommended for bathroom and kitchens because of the many seams, which can collect water and cause the tile to fail.

Resilient tile and carpeting need an underlayment, an extra sheet of floor to make sure the floor is strong enough for people and furniture. If carpeting or tile were secured just to one layer of plywood, you'd get a springy floor, an annoyance at least. So, if the subfloor is ⅝-inch plywood or nominal 1-inch boards, a ½-inch underlayment of plywood or particleboard is required. Nail with underlayment nails, which are ringshanked or spiral. Nail every 6 inches along the edges and every 12 inches on intermediate joists.

Underlayment must be smooth to receive resilient tile. If there are shallow knotholes or other defects, they must be filled with a plasterlike material designed to level floors.

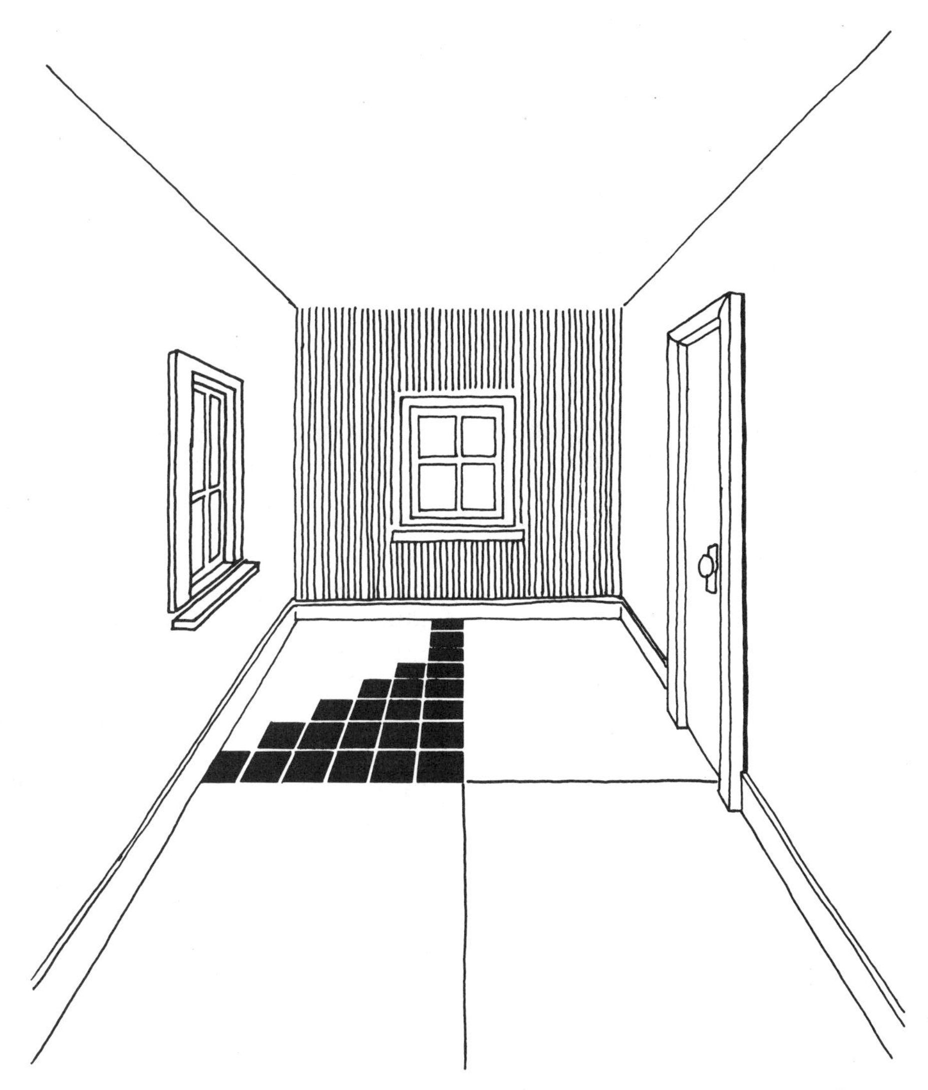

FIGURE 53. To lay resilient tile, or any tile squares, find the center of the room by locating the centers of each wall and connecting the center marks. Start laying tile at the intersection of the two lines.

This is troweled into the holes, smoothed off, and sanded when dry. Joints between underlayment or a plywood floor must be filled the same way.

Resilient tile is made of vinyl-asbestos, pure vinyl, or the "no-wax" vinyl. The vinyl-asbestos is the least expensive and the most available. The variety of designs is remarkable. Some tiles are designed to look like ceramic tiles, with their seams as part of the pattern. Others are designed to disguise the seams or simulate wood. The latter do a good job, but let's face it, they really don't look like wood.

Tiles come in 12 x 12–inch squares that are 1⁄16 inch thick for domestic applications. Commercial tile is a little thicker. Tile is put down either with an adhesive applied to the floor, or with the adhesive already on the back, in which case all you do is peel off the protective backing and press it into place. When using the separate adhesive, use the material recommended for the tile and where it is to be placed (above grade, on grade, or below grade). And follow directions to the letter; otherwise you'll get too much or too little adhesive on the floor. The right amount is critical to a good job.

Because tile comes in squares, it is applied so that each tile is square to itself, not to the room. This is important if the room is slightly out of square (its corners are not 90-degree angles). To do this, find the center of one wall and mark it (Figure 53). Find the center of the opposite wall and mark that point. Then connect the two marks with a snapped chalk line or straight pencil line. Then do the same with the other two walls. You are locating the center of the room. If the room is not square, the lines will not be at right angles to each other (90 degrees). If the room is square, the crossing lines will form four right angles.

The center mark will allow you to start the tiles in the center of the room. It will also allow you to determine the width of the edge tiles, which should be at least half the width of a wide tile. Say the width of the room is 12 feet. The width would be divided into two sides of 6 feet each. Because no room is dead square, you might have trouble making the 1-foot edge tiles reach the edge of the floor. So, move the center line 6 inches to either side. This will divide the width into one section 5½ feet wide and the other 6½ feet, which allows 6-inch border tiles.

The same can be done with a width of, say, 12 feet 4 inches. Instead of a row of 12 tiles and two borders of 2 inches each, moving the center line 6 inches allows a row of 11 tiles (11 feet) and 8 inches for each border tile (totaling 16 inches), which comes out to 12 feet 4 inches. Once you master the technique, it's easy to do it right each time you try it.

Instructions on how to apply the adhesive will be on the can of adhesive. In fact, most packages of tile have instructions. Apply the adhesive with a notched trowel. It's important to use the right-sized notches to make sure the right amount of adhesive is applied.

Line the first tile up on one of the lines (the adhesive is applied right up to the line), right at the center line. If the tile does not butt up on the second line, don't worry; if you follow one line the tiles will be installed the best way possible. On succeeding tiles, butt them up against their neighbors tightly, and drop them into place. They must butt tightly; you can't fudge on the joint as you were able to with certain ceiling tiles. Don't try to push them into place once they're in the adhesive; that will just plow up the adhesive and make it ooze through the joints. If you dropped the tile in the wrong place, pick it up and try again. Tiles with the adhesive already on the back are easier to work with, but once they're down, they're pretty tough to get off. So be right the first time.

Most tile can be cut by scoring it with a utility knife and breaking it at the score mark. To cut edge tile, place the tile to be cut directly on its neighbor, lining it up. Then use a full tile and place it against the wall, overlapping the tile to be cut. Then, using the measuring tile as a straight edge, score the tile and break it. Properly done, the edge tile will fit. If you install baseboards and a quarter-round shoe mold (see Chapter 22), this border tile does not have to be that accurate.

Corner tile is measured and cut the same way, except you have to measure and cut in two directions. On an outside corner, where a closet or any other construction sticks into the room, the tile must be carefully measured and cut. It's a trial-and-error technique.

Resilient tile should not be placed over boards because their seams will "ghost" through the tile and be visible. The underlayment will take care of this little problem, if it is plywood or particleboard. If for any reason the floor on which the tile will go is boards, you can apply ¼-inch plywood or hardboard, or apply roofing felt, securing it with linoleum cement. Don't overlap the edges of the felt; butt them.

Another form of tile is wood, called parquet. It comes in squares of 9 x 9 and 12 x 12 inches. It can be solid or laminated, have a one-piece surface or be made up of a series of strips. It must be sawed to fit at the edges and other areas where it must be cut. It is usually prefinished, and like any other wood, must have a ½-inch gap between edge pieces and the walls. Because it is usually thinner than strip flooring, it cannot be sanded as often to restore its finish.

Wall-to-wall carpeting goes over underlayment, and is best installed by a professional. Some contractors will put down oak flooring even if wall-to-wall carpeting is planned because the carpeting will not outlast the mortgage, and if the house owner ever changes his mind, he'll have a good floor when he removes the carpeting.

For both bathrooms and kitchens, sheet material such as vinyl or linoleum is a good flooring. It comes in widths up to 12 feet, the wider the better to avoid too many seams. Some sheet material is cushioned, which is easy on the foot, but there is this caution: make sure the seams are glued together on installation. If they are not, a weight can press down one edge of the material and a foot can strike the undepressed edge, breaking it.

The no-wax sheet material is among the most modern, and while it does not require waxing for some time, especially the super-shiny stuff, it will eventually need recoating with a material designed just for this type of flooring.

Sheet material usually goes down after kitchen and bathroom cabinets are installed, but before the toilet is installed so the base of the toilet can cover the flooring.

Ceramic tile is a good material for bathrooms, kitchens, recreation rooms, and hallways. It is very hard, of course, and has poor sound-absorbing qualities. It comes in as many or even more styles, shapes and colors than wall tile, ranging from tiny mosaics to huge slabs of unglazed tile including quarry tile. Glazed tile is preferable because it really needs no maintenance except cleaning with a little water and detergent. It is applied with an adhesive the same way as wall tile (see Chapter 20). It used to be set in mortar, but rarely is today, particularly since the mortar adds appreciably to the weight of the floor.

Whether you apply individual pieces of tile or sheets of mosaic held together with paper or plastic mesh, the center of the room must be found in the same way as for laying resilient tile. Depending on the size of the tile, it may not be necessary to adjust the center line to make sure the proper-sized edge tile is applied.

Most floor grout is gray, contains sand, and is applied the same way as the grout in wall tile. It is mixed with water to the consistency of superheavy cream. In fact, if it is a little on the dry side, it is easier to apply. Press it into the joints with a rubber squeegee, making sure all joints are fully and tightly packed.

Remove excess grout with the squeegee, and when the grout starts to set, perhaps in 15 minutes or half an hour, strike the joints with a rounded wood stick or dowel. When the grout has dried on the surface of the tile, remove it with a dry cloth and polish the face of the tile with another dry cloth.

A modern floor covering, designed for bathrooms and kitchens, is a poured or seamless floor. It is applied as a liquid — latex or urethane. Before it dries, colored chips of vinyl or other material are scattered on the surface, and the floor is then covered with a clear glaze. It can be applied to plywood or concrete.

Because different flooring materials have different thicknesses, care must be taken that the thickness of the underlayment is adjusted according to the thickness of the surface material or flooring, to prevent different levels in different rooms.

For instance, wood-strip flooring is $\frac{25}{32}$ inch thick, while most ceramic tile is $\frac{1}{4}$ inch thick, and resilient tile and sheet goods are generally $\frac{1}{16}$ inch thick.

Modern construction dictates that floors between rooms should be on the same level, to prevent tripping. Because strip flooring does not take an underlayment, and is applied directly to the subfloor, a floor of different material in an adjoining room must have an underlayment of a thickness, plus the surface material, to match the strip flooring.

For instance, if the strip flooring is $\frac{25}{32}$ inch thick, a resilient tile floor $\frac{1}{16}$ inch thick must have an underlayment $\frac{1}{16}$ ($\frac{2}{32}$) less than $\frac{25}{32}$, which is $\frac{23}{32}$. Because $\frac{1}{32}$ inch is not critical, increase this to $\frac{24}{32}$, or $\frac{12}{16}$, or $\frac{3}{4}$ inch. So you can use $\frac{3}{4}$-inch underlayment for resilient tile. The same rule applies to ceramic tile or any other material of different thicknesses.

If you use $\frac{5}{8}$-inch underlayment in one room with resilient tile, its floor level will be different from that in a room with strip flooring. This difference can be made up with tiny aluminum beads that can be butted against the resilient tile to prevent breaking of the tile.

Sometimes a marble threshold is used at a bathroom door, but it's not essential. And many old homes have so many different levels that wood thresholds must be installed at the doorways, whether there is a working door or just an opening. Well, that is what building your own house is all about: eliminating these vagaries.

chapter 22

With All the Trimmings

Inside Woodwork

You're coming down the home stretch, and the nearer you come to completion, the longer it seems to take. There are plenty of details left, but essentially you're home free.

After the floor is installed, woodwork — door and window casings and baseboards, etc. — goes in. The floor means a finish wood floor or underlayment, plus tile, but not if you plan wall-to-wall carpeting, which goes on after woodwork.

Windows and exterior doors have been installed in their rough openings, and after walls and ceilings are installed, these doors and windows have to have an interior trim (Figure 54). Doors in partition walls, including openings without doors, also need to be trimmed.

The trim around doors and windows is called casing. It comes in various styles and woods, but the most common wood is pine. Casing is milled to a Colonial or modern look, the latter called ranch or clamshell, which also comes in mahogany wood. Softwood trim, besides pine, is made of spruce, redwood, and gumwood. Very fancy woodwork comes in hardwood, with a very fancy price: oak, ash, birch, maple, and walnut. The hardwoods are to be avoided except as very special accenting, simply because they are so expensive.

You can also use ordinary 1 x 3, 1 x 4, or 1 x 5 pine boards for trim for a simple look. While it is simple looking, it is not necessarily primitive or informal. All trim can be painted or finished naturally with varnish, or stained to the desired color and then varnished.

Casings for both doors and windows can range from 2¼ to 4½ inches wide. Standard thickness is 11/16 inch.

If you have put in setup windows, which include everything but the inside trim, you must have jamb width to fit the wall, critical to proper installation of inside trim. Standard width is 4½ inches, for a wall totaling 4½ inches. A 4½-inch wall is made up of 3½ inches of a 2 x 4 stud, plus ½ inch for exterior sheathing and ½ inch for interior plasterboard. A 5½-inch jamb is necessary for a 5½-inch wall. The extra inch includes a plastered interior wall with a plasterboard lath. Other wall thicknesses are determined by the thickness of sheathing, which can be ¾-inch boards. The most important thing is to make sure the thickness of the jamb corresponds to the thickness of the wall.

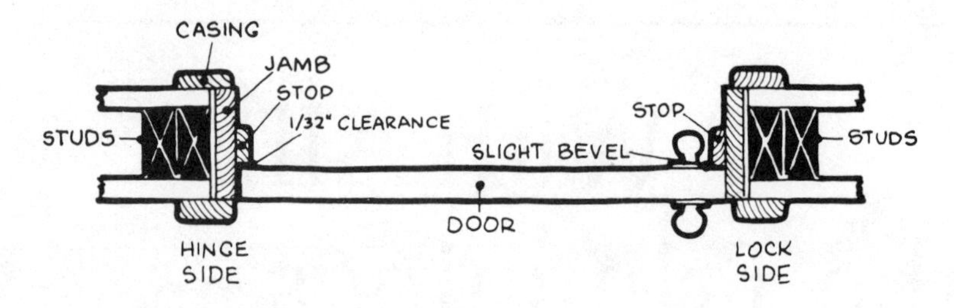

FIGURE 54. Jamb and casing for a door, which is typical interior trim for both windows and doors, including openings that have no doors.

The first trim to go on the interior side of the window is the stool (inside sill), which is notched to fit over the sill and butt up against the bottom sash. The stool should be notched at each end to clear the window opening and butt up against the finished interior wall. Its length should extend ¾ to 1 inch beyond the outside casing edge. Use 7d or 8d finishing nails to install.

Next, the top and side casing is nailed so it is flush with the edge of the jamb. It is nailed to the rough opening studs (through the inside wall) and to the edge of the jamb. Use 8d finishing nails. Milled or molded casing must be mitered at the two top corners; otherwise its thinner edge would jut out in a most unprofessional way. The miter is at a 45-degree angle and the cut is made with a miter saw in a miter box. Plain casing like a 1 x 4 or 1 x 5 can be butted in a straight joint, with the top casing resting on top of the side casing. In a butt joint, the ends of the top casing will show, and because they are rougher than the side or flat face of the wood, they are hard to paint, and stain will be much darker than on the face. One Colonial technique is to allow the top casing to overlap the side casing by ¼ inch on each side. This is a matter of taste.

Next come the side and top stops, which range from 1⅜ inches for the sides to 4½ for the top. They can meet the edge of the casing or come within ⅛ or ½ inch of the edge. Nail with 6d finishing nails. All nailing of casing is done in pairs, approximately 16 inches apart.

Finally, the apron is installed under the stool, its length equaling the width of the casing, outside to outside. Drive 8d nails through the stool into the apron.

Other types of windows call for a different trim treatment. Awning and hopper windows usually don't have a stool but a casing at the bottom. Casement windows can have either treatment.

Interior doors are cased the same way as windows. Doors come set up, with one side of the casing installed, so they can be slipped into rough openings. Again, the thickness of the jambs must correspond to the thickness of the wall.

Doors come in several styles: paneled, flush, and louvered. Flush doors are the least expensive, although if they have birch veneers they can be finished natural or stained and varnished for a rich appearance.

Door widths vary, but interior doors are usually a maximum of 32 inches, and range down to 2 feet or less for closets. Standard heights are 80 and 78 inches. While exterior doors are usually 1¾ inches thick, interior doors are 1⅜ inches thick. Louvered doors, for appearance and for ventilation, are 1⅛ inches thick.

A setup door is installed in a rough opening with ¼- to ½-inch clearance on both sides and top. The jamb has ears on the top, projections of the side jambs, so there is no need to nail it at the top. One side jamb is plumbed in the rough opening and the top is leveled. Then, shims (shingles that taper) are driven between jamb and jack stud until the jamb is tight, and are nailed.

If the gap is extra wide, two shingles, with narrow ends sliding over each other, are driven in. As each shingle is driven in, from opposite sides, they'll tighten up and fill the gap. Be careful not to drive them in too far; they can bow out the jamb very easily. Nail with 8d finishing nails, in pairs, every 16 inches. One of the paired nails can be positioned so the door stop will cover it. The shingles should be placed so that the nails will be driven through them. Use plenty of shingles; too few can cause a wavering of the jamb. Then, plumb and shim the other jamb.

Make sure the door fits properly, then put on the casing, which is installed the same way as window casing, except that the inner edge of the casing is indented $\frac{3}{16}$ to $\frac{1}{4}$ inch from the edge of the jamb to give a finished look. You can even make the indentation $\frac{3}{8}$ inch; the bigger the indentation the more detail you'll get. You can use a strip of wood or hardboard of the right thickness as a guide for the indentation.

Nail with 8d finishing nails through the thickest part of the casing (if the casing is tapered); use 5 or 6d finishing nails through the narrow part, which goes next to the jamb. If you use a regular board of equal thickness for casing, use 8d finishing nails. Again, nail in pairs and space them 16 inches apart.

If you don't use or can't get setup doors, the work must be very precise so the door will fit properly. Sometimes you can buy custom-made jambs, which help. Otherwise they must be made in place. So here is where dimensions must be precise. Jambs must be installed so the clearance between door and jambs is as follows: $\frac{1}{8}$ inch at the top, $\frac{1}{8}$ inch at the latch side, and $\frac{1}{16}$ at the hinge side. The floor clearance is $\frac{1}{4}$ to $\frac{1}{2}$ inch, with more necessary for a carpeted floor.

Placement of hardware is arbitrary, but generally the knob goes 36 inches from the finish floor and hinges are placed so the bottom of the lower hinge is 11 inches from the door bottom and the top of the upper hinge is 7 inches from the door top.

Some jambs, particularly for exterior doors, are made with the stop (to hold the door closed in the right position) all in one piece. Some interior doors are made this way too, and some jambs have two interconnecting pieces so they can be opened to accommodate the thickness of the wall.

But when doorstops are separate, they are nailed in position temporarily to hold the door in the right position for the installation of hardware.

Before fitting the door, put the hinges on the door edge. Interior doors take $3\frac{1}{2}$-inch hinges, called butts, with loose pins. Exterior doors take two 4-inch hinges, with a third in the center of the door edge to add strength and prevent warping. Remove the pin and lay half the hinge, or one leaf, in position on the edge of the door. The gap between hinge and edge should be at least $\frac{3}{16}$ inch. The hinges should be mortised (embedded) in the door edge and casing so their surfaces are flush with the wood surface.

If you don't have a router for this purpose lay the hinge on the edge of the door, making sure it is square. You want the hinge to fit snugly in the mortise, so instead of making a pencil line around it, cut around it with a utility knife. And to keep the hinge in place when you're cutting, mark the screw holes, predrill them, and drive at least two screws to tighten the hinge. Then you can cut hard and deep enough with the utility knife around the hinge, making the mortise a perfect fit.

When you've cut around all three sides of the hinge, remove the hinge and chisel out the mortise with a sharp chisel. Sharp is the important thing; otherwise you can gouge the wood and the mortise will be useless. You'll have to check the fit of the hinge so that it will be flush with the wood surface; not too deep and not too shallow.

With the hinges on the door edge, position the door in place with shims to make the correct gap at the bottom and the stops in place for vertical alignment. Then mark the position of the other leaf of the hinge. You can line it up on the leaf already on the door edge or put the second leaf in place with the pin. Then, follow the same technique in mortising the jamb for the second leaf of each hinge.

There are no-mortise hinges on the market, which are hollowed out or made of a series of "arms" instead of a full leaf, so that one leaf snugs into the other. These are not as strong, satisfactory, or attractive as the mortise hinges.

Interior latches include simple latches that cannot be locked, a lockable latch for bedrooms, if desired, and a bathroom latch, which is lockable from the inside with a slot or hole in the knob that receives a pin for unlocking in the case of an emergency.

Instructions and templates for position come with lock sets. Most inside latches are fairly simple, and require holes in the face of the door for knobs and a hole in the edge for latch, lock, and faceplate. Exterior locks are more complicated, and the good ones are equipped with outside key, inside lever with a dead bolt for jimmy-proof security, and a night latch.

Many setup doors don't have very good locks on them, so when you buy setup exterior doors, make sure the locks have a dead bolt, which is nonbeveled, fits into a striker plate (in the jamb), and is difficult to jimmy without tearing the door or jamb apart. Some exterior locks simply make the beveled latch nonmovable, but unless there is a security pin it can be moved easily with a plastic credit card. Of course, locks are made for honest people, and very few locks cannot be opened by a determined burglar.

The striker plate is mortised into the jamb so that the beveled latch catches in the opening. In the case of exterior doors, the plate has an opening for both beveled latch and deadbolt. Mortising to take the striker plate is done the same way as for hinges. Carve the opening for the latch deeper so the latch can fit into it without interference.

When everything is installed, the door should swing freely without closing or opening of its own accord. If the door swings one way or another, the upper or lower hinges can be set a little deeper to check the swing. Finally, the doorstop is installed permanently in its correct position. The stop along the latch side is installed first, butted snugly against the face of the door when it is latched, so there will be a minimum of play between latch and striker hole.

Nail the stop with 5 or 6d finishing nails,

pairs spaced 16 inches apart. Then put up the top and hinge side stops. If the stop is molded, the corners at the top must be mitered at a 45-degree angle. If the stop is plain, the top stop can be the full width of the jamb with the side stops butting against its under side. The hinge side stop should have a 1/32-inch clearance between stop and door so the swinging door edge will clear the stop.

One practical aspect of door hanging concerns the direction in which it swings. Doors should swing into rooms in the direction of natural entry; for instance, they should swing into bedrooms that have no other exit or entrance. Doors should not swing into hallways, which are narrow at best, and should not interfere with other swinging doors. They also should not swing into the stairways, which can be dangerous. You usually have to compromise at a stairway, since a door swinging away from it usually has to swing into a hall. Well, you can't win all the logical arguments. Exterior doors always swing in.

The only other molding in a house, generally, is base molding, often called baseboard or mop board (Figure 52). A baseboard is important to act as a bumper against furniture and other abuse and, in a practical sense, to protect plaster or plasterboard walls from the wet mop being sloshed around when floors are washed. You never use a wet mop (only damp, well squeezed out), of course, on wooden floors, but even so, waxing machines and vacuum cleaners can raise hell with a wall.

A baseboard can be a plain board, 1 x 4 or wider (old houses have them up to 12 inches or so). A plain board is naked without a base cap or top molding, often called a band, which is applied to the top of the baseboard. It can also be one-piece molding.

Nail the baseboard through the wall into the studs. Use 8 or 10d finish nails, a pair to each stud. Smaller nails can be used to nail on the base cap, but they should be nailed into the studs also. Plain boards can be butted in inside corners and mitered on outside corners. Any molded baseboard must be mitered at both inside and outside corners.

In conjunction with the baseboard, a shoe mold is installed at the floor line. A true shoe mold is a quarter round (a quarter of a dowel) with one edge 1/2 inch and the other 3/4 inch, with the wide edge nailed against the baseboard. A regular quarter round also can be used. A shoe mold is not necessary, but makes a nice trim and keeps furniture from hitting baseboard and wall. It should not be used with wall-to-wall carpeting.

The shoe mold should be nailed directly to the baseboard, and not to the floor. This is to prevent the floor, when it contracts due to moisture loss, from pulling the shoe mold away from the baseboard, creating a gap. The shrinkage of the baseboard is minimal, so only a very small gap between shoe mold and floor would result in any case of shrinkage.

Sometimes a piece of molding, called a crown, is applied to the wall where it meets the ceiling to cover a bad joint or to avoid the need for taping and covering the joint with joint compound. Some purists will pooh-pooh this as lazy, but ceiling molding is very decorative, and one type, picture molding, is highly prized in old homes. Crown molding is usually nailed to wall studs and ceiling joists with 8d or 10d finishing nails. Crown molding can be small or large, but it is usually hollow inside.

Because it is quite elaborate, crown molding should be cut with a coping saw. This technique is preferable to mitering with any molding if the corners are not square or the molding elaborate. The idea is that instead of mitering the joint, the end of one molding is cut to follow the contours of the other molding when it butts against it. So, install one piece of molding so that it butts flush against the corner. Then, miter the end of the other molding at a 45-degree angle. Then, with a thin-bladed coping saw, square-cut the end of the molding, following the contours created by the miter cut. With a little practice, it works well.

There is a new molding material called cellular vinyl, which comes in a natural "wood" color and an oak embossing. It can be stained with pigmented stains and is said to resist burning, be capable of being sawed and nailed, and be less expensive than wood. It is not like the earlier plastic moldings, which are made of PVC (polyvinyl chloride) and had a photographic wood grain surface laminated to the plastic.

A word about finish nailing and countersinking. If you use finishing nails, don't try to drive them flush with the wood surface. You'll simply mar the wood and be called a butcher by finish carpenters. And they'll be right. So to avoid this, simply stop nailing before the head is flush with the wood. Finish the job by setting the nail head below the surface. If you're doing a lot of finish nailing, buy nail sets of different sizes. Too large or too small a nailset is hardly better than no nailset at all.

In a pinch, you can countersink a finishing nail with the point of a larger nail. Also, you can lay a large common nail (with a big head) on the wood so the edge of the head touches the head of the finish nail. A judicious hit or two with the hammer will countersink the finishing nail's head.

Once the nails are set, fill them with wood putty, a powder you mix with water. Sand the putty smooth, then paint. If you want a natural finish, the color of the putty should be fine. For a darker stain, mix the putty with a water-soluble stain, details of which are in Chapter 26.

If you plan to finish the wood with a natural tone, or a stain, avoid the use of solvent-thinned putties; they tend to stain the wood and are difficult for stains to penetrate, varnish to cover, or sandpaper to remove.

Real finish carpenters, particularly when working with furniture, will cut out a little sliver of wood where they plan to drive a nail, and when the nailhead is set, glue the sliver back over the head. And to do a really good job, some carpenters will simply lift part of the sliver off, with one end acting as a hinge, and put it back and glue it when the nail is driven and set.

Putty will do very well for most of us.

chapter 23

Everything in Its Place

Cabinetry and Other Millwork

Everyone who has ever daydreamed has probably conceived of the ideal kitchen in which everything has its place, all work is so efficient that the kitchen worker saves a bundle of time, and everything is beautiful.

Dreams don't cost any money, but kitchens do. They have the most millwork in the house. Millwork refers to manufactured, built-in components, such as cabinets, shelves, china cabinets, fireplace mantels, and small closets.

Design can sometimes lead the dreamer astray. After World War II, when vast housing projects went up all over the country, the designers did something that proved to be unfortunate: they designed efficiency kitchens where the appliances were tucked away in an efficient little part of the apartment, with some kind of dining alcove off the kitchen but not really a part of the kitchen. Well, many people have for generations virtually lived in the kitchen. Everything was done around the big kitchen table: bills were figured, homework was done, family problems ironed out, courting carried on. You name it, the kitchen table was the focus of the family. While this was going on, Mama presided, or was there to give aid and comfort. She had plenty to do, but in the big old kitchens, she was near or at the kitchen table while she went about her chores. Well, those big, efficient projects isolated the kitchen, and Mama, from the table, which was now in an alcove or "dining area," which might not have been the best idea in the world.

So it's important to design a kitchen to fit family needs, desires, and dreams.

You have to customize your kitchen generally but only generally, to meet your needs, which will change over the years of occupancy. We say generally, because if you overdesign; that is, design to too close a specification for your needs and tastes, you might find it hard to sell the house when you're ready to do so.

This restriction also applies to the number and location of bookcases, china cabinets, utility closets, and other little "extras." For instance, if you are a real collector and reader of books, you're going to want to have lots of book space. The next occupant of the house might not find books his bag, and will find bookcases a waste of space. They do take up valuable wall space. For instance, many designs incorporate bookcases, indeed, entire storage walls, on each side of the fireplace. This is fine, but it can pose a problem of furniture arranging.

These are the pitfalls of designing your own house. So let the builder beware.

A builder can build his own kitchen cabinets, which is fine when he knows what he is doing and what he wants. But there are so many cabinets on the market today that it probably would pay to buy components and install them.

The least expensive types are knocked-down pine or fir cabinets that the builder puts together and paints or stains to his taste. These are good cabinets. With stain or paint and the right hardware, they can be cheerful, elegant, or any other adjective you want to choose. The more elaborate cabinets are made of hardwod (maple, birch, oak, etc.) and are prefinished in a variety of stains and designs. The fancier you get, the higher the price.

Even before choosing cabinets, you must decide what kind of a kitchen layout you want (Figure 55). There are four classic arrangements: the U shape, L shape, parallel walls, and one wall. They all contain the basics of a kitchen: stove, refrigerator, and sink, and sometimes a dishwasher. But they should also contain an eating area as an integral part of the kitchen, essential when there is no dining room, and a good adjunct to a house even with a dining room.

Today's kitchens are loaded with cabinets, both base (on the floor) and wall. They have lots of nice shelves and doors, but basically they're pretty inefficient. So, if at all possible, provide room for a pantry in the design of the house. Whether large or small, a pantry should be arranged with shelves on each side of the room, perhaps with enough space for clothes washer and dryer. Open shelves are readily accessible, and if the pantry is not in full view of kitchen and/or dining room, the things on those open shelves don't have to be terribly neat. For practical purposes, the pantry should be as close as possible to the eating area in the kitchen and to the dining room.

Lower shelves should be at least 18 inches deep for large pots and pans. Upper shelves should be at least 12 inches deep, or 11¼ inches, which is a nominal 12 inches for a 1 x 12. One area can be set aside with dividers for all those long and wide cookie sheets, shallow pans, and other flatware. Shelves can be equipped with cup hooks to save space. Also included in the kitchen could be a broom closet, and finally, a built-in ironing board cabinet. This latter was considered obsolete when the modern "portable" boards were developed. But the portable boards are very heavy, and tend to be left out in any room the ironer wants to iron in. With a cabinet, the board goes down and up with ease, and is really out of the way. And with a big enough kitchen, the ironer won't feel left out of the family action. Of course, pantry and laundry can be large or small, but it is possible to put a lot of storage and working appliances in an area as small as 6 x 8 feet. This space could also be added to for a downstairs bathroom, either half or full.

Broom closets and ironing cabinets come ready-made, or you can build your own. An ironing cabinet is simple; if you can't find a wooden board to go into a cabinet, build one out of ¾-inch plywood, with one smooth side.

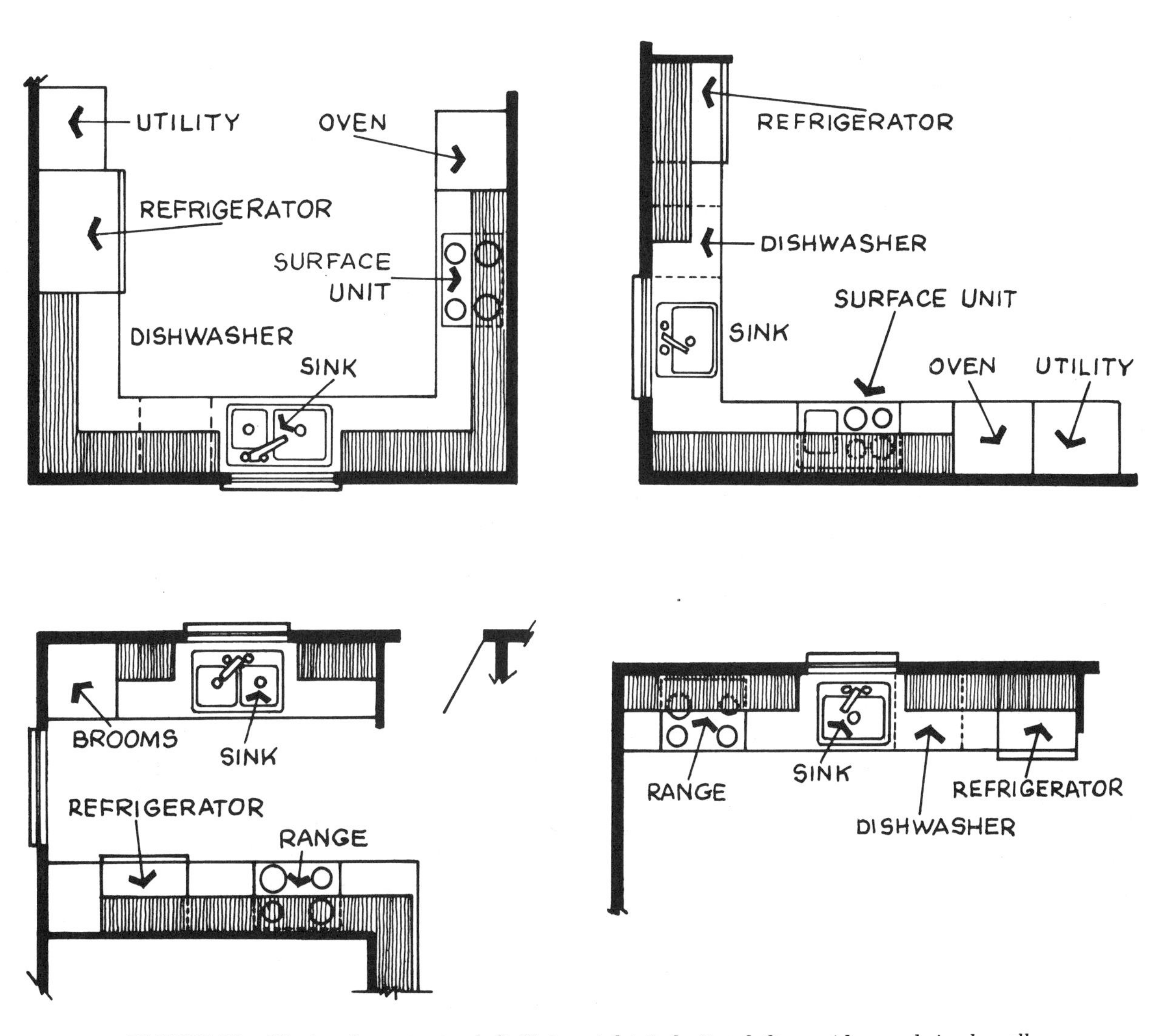

FIGURE 55. Kitchen layouts: top left, U; top right, L; bottom left, corridor; and single wall.

FIGURE 56. Popular cabinet style at left; style at right is considered better for convenient storage. Dimensions remain the same for both. Even better, at least in conjunction with cabinets, is a separate pantry.

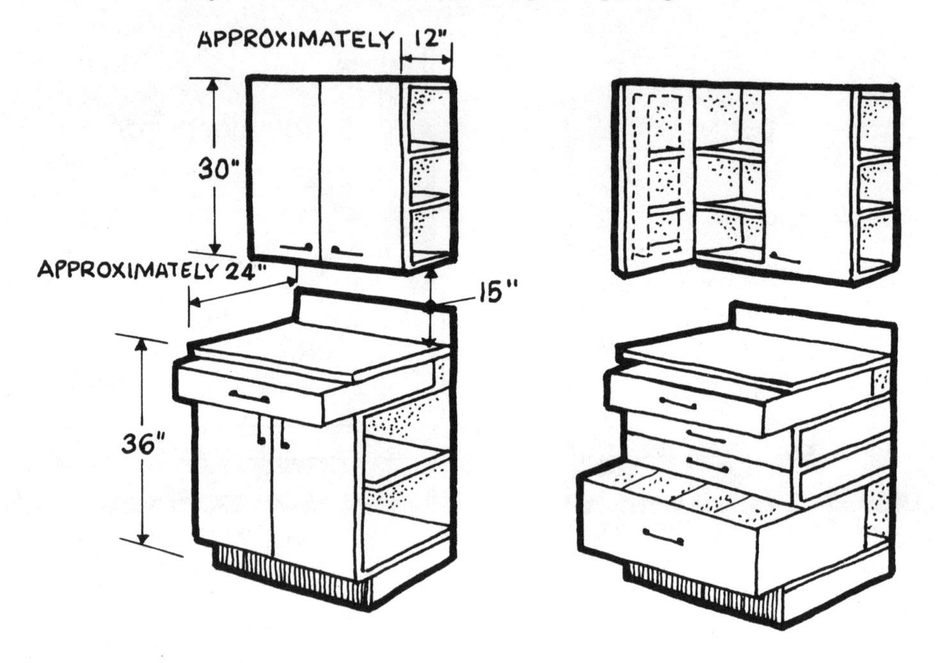

Trace the old "portable" onto a sheet of plywood and cut it out. Two T-hinges can hold the board on the wall (the long tongue of the T-hinge attached to the board, the smaller leaf to the wall), and a hinged leg can be attached to the narrow end of the board for stability.

An old-fashioned screen-door spring can be used to retract the leg, and an ordinary wooden latch, secured by a screw, can keep the board from coming down on your head when you open the cabinet. The cabinet can be built right into the wall, between 2 x 4 studs, and will stick out very little if at all into the room. The ready-made cabinets are designed to fit between studs.

As for the kitchen itself, the cabinets do play an important part in kitchen work (Figure 56). The base cabinets provide for a working surface that takes the place of the kitchen table, whose height caused a lot of backaches in the "good old days."

Base cabinets are generally 36 inches high and 24 inches deep. Their width depends on the size of the kitchen and the number of appliances in it. It's not always convenient to store things deep in a 24-inch-deep cabinet or to put other things in front of them. A base cabinet is good for storing flatware like baking sheets and large pans in a divided compartment. But for most purposes, drawers are better. The general recommendation for the amount of base cabinets ranges from a minimum of 6 feet to a maximum of 10 feet. If, in a corner, a lazy susan circular shelf is installed, it helps to increase the storage to the recommended amount.

With 24-inch-deep cabinets, usually with

an indented area (kickboard) at the floor to prevent toe-stubbing, you get plenty of counter surface for work. This counter surface is particularly important next to sink, stove, and refrigerator. Experts have come up with minimum and maximum amounts of counter next to appliances: refrigerator, 15 to 18 inches; sink, 24 to 36 on one side for stacking dishes, 18 to 30 inches on the other side for clean dishes. If a dishwasher is installed, these figures can be reduced, and the 24 inches or so of counter space over a built-in front-loading dishwasher is so much bonus. For mixing and preparation, 36 to 42 inches; beside the stove, but not including range surface, 15 to 24 inches; and beside the oven, if it is separate, 15 to 18 inches.

Then you come to wall cabinets. These are generally 12 inches deep, and their bottoms should be a minimum of 15 inches above the counter, and a minimum of 24 inches above stove and sink. Their tops should not go higher than 7 feet above the floor. If there is space between top of cabinets and ceiling, it can be left open or covered with a facia, which is just a blank wall. This space could also be filled with doored cabinets to store off-season or rarely used supplies and utensils. However, if this space is used for storage, a good, sturdy ladder stool should be provided.

One word about cabinets. We have pointed out the disadvantages to some kinds of base and wall cabinets, particularly deep ones. So, if there is a wall, say, which is one side of a passageway, and there is no room for a deep cabinet or furniture, use your imagination and build a glass cabinet. Not one made of glass but one for glasses: it can be just 3½ inches (the depth of a 1 x 4 board for the shelves and the width of a 2 x 4 stud), so it can be built right into the wall.

Building into the wall is no good on an exterior wall because there would be no room for insulation. But on an interior wall it could work nicely; either with the studs in place or taken out, with a sheet of ¾-inch plywood taking the place of the plaster or plasterboard on the other side of the wall.

Anyway, you'd be amazed at the number of glasses, bottles, and canned goods of various sizes and shapes that can be stored in a space 6 feet wide and 7 or 8 feet high. Put doors on it and you have convenience, because the shelves are so shallow that you can't get more than one row of anything on them.

Some of the modern cabinets have very well-organized shelves, including some on the cabinet door, and they serve a purpose similar to that of the glass cabinet. Adjustable shelves also allow versatility in storage, and once you get the storage area organized, you can adjust the shelf height to save space.

Most cabinets are prebuilt in almost any dimension, so you can gang them into almost any length possible. They have front and rear frames for securing to the studs in the walls. Some installations require or recommend that furring strips be nailed to the studs first; then the screws can be driven into the furring strips and their horizontal position is not critical.

Frames are also made so that the cabinets can be connected to each other and to the ceiling, if necessary. It is most important to make sure the cabinets are level, both along their width and along their depth. If the floor is not level, shim the cabinets so they will be level by using shingles under the base frames.

Cabinets also come with accessories like base and wall fillers, for filling spaces too narrow for a cabinet but necessary to bring the cabinet out to the right length. There are fillers for ovens, ranges, and sinks.

Then there are special accessories like a pull-out desk top; spice cabinets (similar to the glass cabinet), lazy susans for corners; bread drawer cabinet with a metal-lined box and a pull-out hardwood cutting board; vegetable drawer bins lined with metal; a heavy pull-out chopping block of hardwood with a groove for liquids.

Most counter tops are made of laminated plastic and modern cabinets are faced with wood grain or plain-colored laminated plastic. You can also make a counter top, including splashboard and stove back of ceramic tile, either the standard 4¼ x 4¼–inch tile or the mosaic type.

Another good counter top is the so-called butcher block, strips of hardwood (usually maple) laminated to each other. This provides a 100 percent cutting surface. To prepare the surface, remove the wax, if any, with mineral spirits and treat it with mineral oil, not linseed oil. Just slop on the mineral oil and let it sit for 15 minutes, then wipe it all off; otherwise it will remain sticky. Give the wood two coats; then coat it once or twice a year for maintenance. Linseed oil is not recommended because it can be toxic.

Finally, there are the manufactured marble counter tops, and tops of acrylic that are also cutting boards and are claimed not only to stand up under wear and tear but also not to dull knives.

Other ready-made millwork items are fireplace mantels and china cabinets. Fireplace mantels can be modern and simple, and often without much of a mantel at all, or Colonial, in several styles, or elaborate, usually custom made. However, if you were to investigate a large wrecking company, you might be surprised to see many fine old mantels that have been taken, intact, from old houses. They are really sort of exciting, if such ornateness turns you on. They are often of pine or hardwoods like mahogany or oak. You can remove the paint, varnish, or any other finish and stain and varnish them to your taste.

China cabinets can be built into corners, or along a wall in the dining room. Sometimes the cabinets are in a wall dividing the kitchen from the dining room and have pass-through spaces at counter height. And sometimes these cabinets have doors on both sides.

Built-ins sometimes can restrict dining room arrangements, so think twice about them. Without built-ins, you can buy new or used or antique cabinets that can be put in any part of the dining room, or, for that matter, any part of the house.

Another "non–built-in" is a wardrobe or armoire, a large cabinet designed for holding clothes that hang from poles or are stored in drawers. Such cabinets are available new, used and antique. Speaking of antiques, or very old furniture, they can do very little but increase in value.

Built-ins can be installed before finish flooring is put down, particularly in kitchens. For china cabinets it is a good idea to put down finish floors like oak-strip flooring first, then build the cabinet on the floor. With wall-to-wall carpeting, everything is put in first.

chapter 24

Ups and Downs

Stairways

Stairs are another built-in item in the house, and millwork manufacturers have many kits designed in early American, Colonial, and contemporary styles (Figure 57).

Stair kits come with proper treads, risers, balusters (spindles), newel posts, hand rails, starting steps, rosettes (flat decorative plaques for connecting a rail to a wall), newel caps, and step brackets. They're all hardwood, usually birch for everything but the treads and risers, which are oak. What you have to provide are the stair carriages (notched beams on which treads and risers are nailed, often called stringers), plus stringers, which are the baseboards that go along the steps against a wall, and possibly molding to support the tread where it overhangs the riser.

Before you get all this stuff, you must build an opening for the stairs, whether it's from the basement to the first floor or first floor to the second floor. And then there are attic steps, if necessary.

The openings for stairs must be built in each floor when the floor is framed (see Chapter 8 and Figure 15). An opening for a basement stairs must be at least 9½ feet long x 32 inches wide. The long dimension can be parallel to the joists or at right angles to them, but it is most desirable to make it parallel to the joists. Stringer and header joists have to be doubled to support the weight of the stairs.

Oak is the best material to use for treads, even if they are to be carpeted. In basement stairs, softwoods like hard pine or fir can be used, but in both cases, enclosed steps (treads plus risers) are a good idea to prevent the area under the stairs from collecting the dirt and debris that open stairways produce.

Width of the stairs, height of the riser and depth of the tread, and headroom are all-important considerations (Figure 58).

For basement stairs, width should be a minimum of 2½ feet. Head clearance, that is, the height from each tread to the ceiling directly above it, must be a minimum of 6 feet 4 inches.

For the main staircase, width should be a minimum of 2 feet 8 inches clear of any part of the stairs, and 3 feet is better. Head clearance must be a minimum of 6 feet 8 inches, to prevent heads suffering from a bang on the overhead.

Tread depth and riser height are critical to comfort in climbing and descending stairs, and are in ratio to each other according to this formula: the depth of the tread plus

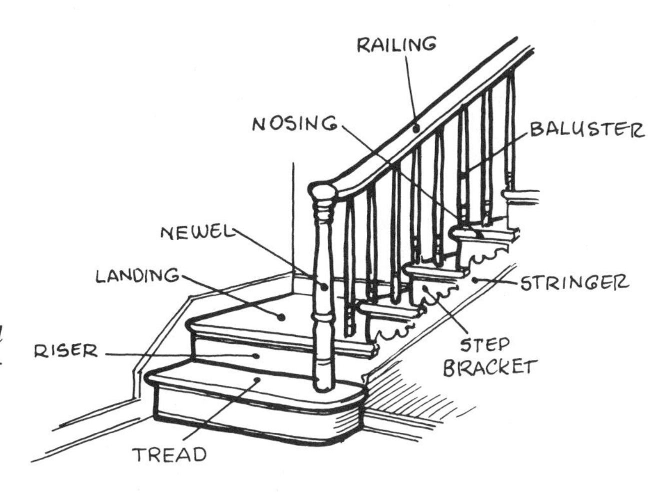

FIGURE 57. Stairway parts. This is a traditional design, but there are many other designs available in kit form.

twice the height of a riser should equal 25 inches. A good riser height is 7½ inches; twice that is 15, so the tread depth should be 10 inches (Figure 58).

A higher riser height can fatigue the climber unnecessarily; too low a height can put the climber off stride. If a higher or lower riser is necessary, depth of the tread should be adjusted to fit the 25-inch formula. For instance, if the riser must be 8 inches, twice 8 is 16, so the tread should be 9 inches deep. If the riser must be 6½ inches, twice that is 13, so the tread depth should be 12 inches. The formula has been designed to be consistent with the stride of the climber.

The amount that the tread overhangs the riser is about 1⅛ inches, and is called a nosing, which requires a trim below it to support it. In determining the height of the riser, the recommended height of 7½ inches includes the thickness of the tread. The depth of the tread runs from nosing to nosing.

Fourteen risers are standard for an 8-foot ceiling. The 8-foot ceiling, plus joist depth and floors, totals about 105 inches. Divide 14 into 105 and you get 7½ inches, which should be no trouble at all to install. If the ceiling and joist height is less, the 14 risers average out to less than 7½ inches, which is fine, as long as you adjust the depth of the tread to the 25-inch formula. And considerably less ceiling height (97½ inches including joist and floor thickness) would require just 13 risers.

There are several types of stairs: straight, L-shaped, and U-shaped, plus variations. If you have the room, a straight stairway is least expensive and easiest to install. The other types will fit into less space. But they require landings, which should be a minimum of 2 feet 6 inches deep, to allow for turning space and to accommodate in-swinging doors; otherwise the door would swing into an open stairway, very dangerous for even the most agile stair climber.

Landing frames are installed, usually in corners, by nailing a 2 x 8 cleat onto the studs, with the outer corner supported by a vertical post of a 4 x 4 or two 2 x 4s spiked together.

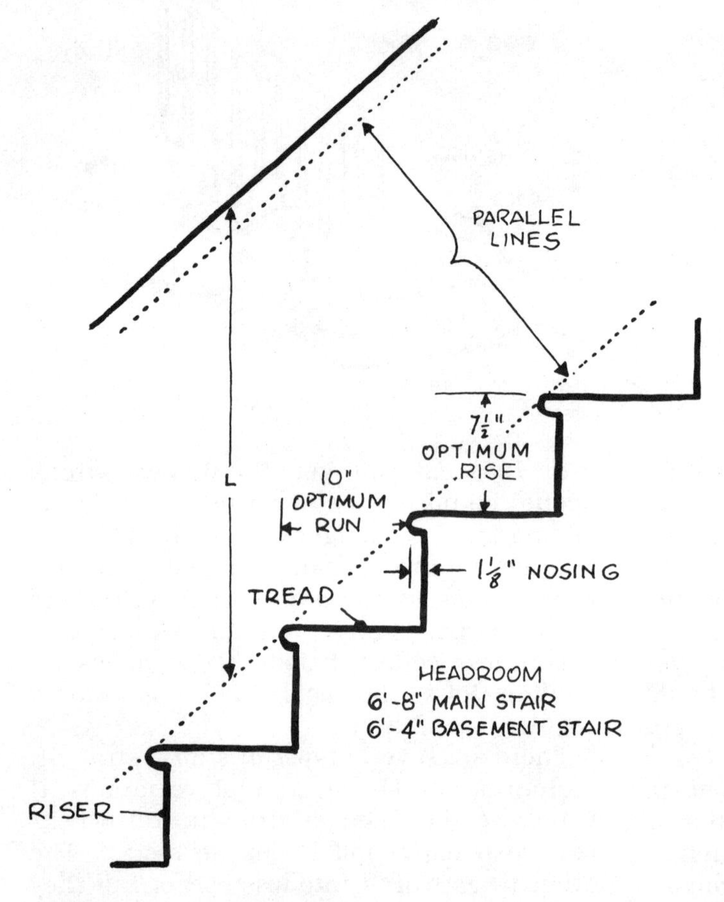

FIGURE 58. Standard dimensions for steps, and standard clearances for stairways.

Sometimes there is no room for a landing where the staircase turns a corner, and "winder" treads are used, often called pie-wedge treads, which taper from an extra-deep tread dimension to nothing, allowing three treads to turn the corner. These treads are not allowed by some building codes, and if they are, there must be at least 10 inches of tread depth at the center of each tread, so that the stair climber, who usually steps on the center portion of the tread, will have enough tread to support his feet.

Handrails must be on the side of an open stairway, and on both sides if the stairway is open on both sides. The double open stairway occurs usually in basement stairs. A basement rail can be a 2 x 4 planed and sanded smooth to prevent slivers, supported by a 4 x 4 or 2 x 4 secured at the top and bottom of the stairs. Sometimes the top of the rail can be secured to the wall. The posts are secured by bolts and nuts along the side of the stair carriage so that they won't work loose. Finish stairs are equipped with hardwood rails, the pieces of which are connected by a special screw and bolt system that is concealed in a groove at the bottom of the rail. This bottom groove, sometimes just a round hole, is covered with a wood plug.

Some stairs are made with a housed stringer, which is a 1 x 12 board routed half its depth to take the tread and riser, which are grooved and rabbeted to each other to increase strength and durability. Treads and risers are wedged, nailed, and glued.

All nailing is done with finish nails. The treads are nailed into the stair carriage and risers, and the risers are also nailed from the back of the staircase and into the edge of each tread.

Stair carriages are usually 2 x 12s, notched to receive treads and risers. There must be at least 3½ inches of solid wood from the deepest part of the notch to the back of the carriage. The thickness and width of the treads will dictate how many carriages to use. Three carriages are needed with ¾-inch treads more than 2½ feet wide, and 1½-inch treads more than 3 feet wide.

In finishing off a staircase, the finish stringer (baseboard following the contour of the stairs) can be placed behind the notched stair carriage and the treads and risers butted against the stringer. Another technique is to notch the stringer in the opposite direction of the carriage so it will fit right against the carriage and the treads and risers butt against it. The former technique is easier, and gives the treads and risers more nailing surface into the notched carriage at the ends of the treads and risers.

If the attic does not have a regular staircase, usually enclosed by two walls, a folding staircase can be installed in an opening in the attic floor. The opening can be as small as 26 x 54 inches. Folding stairs come already constructed and ready to put in.

Staircases don't have to be of traditional design. They can be ultramodern, with, for instance, a special center carriage with treads extending on each side, without any other support, or can be "flying," attached on one side only (cantilevered). Sometimes they don't have risers, giving them an airy appearance. But no matter what their design, they should follow the 25-inch rule for ease of going up and down.

Finally, there can be spiral staircases, made of metal or wood, or a combination of both. Spiral staircases come ready to be installed, or in kit form. There should be a straight staircase in the house somewhere, however, to accommodate furniture moving.

Outside stairs are made the same way as basement stairs. They can have regular treads and risers, or they can have treads of 1 x 3s spaced ¼ inch apart for drainage. Or they can be nominal 2-inch lumber, say two 2 x 6s spaced ¼ or ½ inch. Remember, no matter what the tread thickness, that thickness must be incorporated into the riser height.

All wood materials on outside stairs should be treated with a wood preservative, and the bottom of the stringer should be based on a concrete footing, preferably sunk below the frost line and topped with a foundation.

Sometimes cedar posts, treated with a preservative such as creosote, are sunk in the ground and stringers set on their ends. This is also done for small porches, too, but the posts will not last as long as concrete.

chapter 25

Things That Go Bump in the Wood

Preserve Your House

The Old Scottish prayer begs for protection against goblins and long-leggety beasties and things that go bump in the night. Once you get all that wood together in your house, there is a host of insects and plants that would like to turn the house into a meal, or their own house, and turn your dream house into a nightmare.

The insects are things like termites, carpenter ants, powder post beetles, and the old house borer. The plants are fungi, some that just stain the wood, others that decay it.

So, before much construction goes on, learn the techniques of construction and treatment to prevent this damage or at least reduce its likelihood.

First, all about the culprits and what they can do.

Subterranean termites occur in most of the United States except the extreme Northeast, northern Michigan, the northern plains, and a very small part of the Northwest. They occur frequently enough to require proper precautions. Subterranean termites live in the wood and eat it and other cellulose material, but they have to go to the earth at least once a day to obtain moisture.

They attack wood in contact with the ground, and if there is no wood in contact with the ground, they can build small tunnels, about ¼ to ½ inch wide, of mud, along foundation walls to protect themselves going to and from the wood. They eat wood fairly close to its surface and can be detected by tapping the wood. If the wood sounds hollow, or breaks up, chances are that it is infested with termites.

Dry-wood termites are common in the tropics, and in the United States are so far found in a narrow strip along the southern states, from Virginia to northern California. They are harder to control, but can be detected by the "sawdust" they leave.

A third termite, the Formosan subterranean termite, has appeared in localized southern areas, and has caused considerable damage. Methods of control for it are similar to those for subterranean termites.

Sometimes in the spring and summer, and even in the fall, you might see a lot of flying termites. They are swarming, which means that they're breeding and looking to set up new colonies. If you see swarming termites in your yard, it doesn't mean that your house is infested, but it's a good idea to have an exterminator or pest control expert check it.

Termites are distinguishable by their thick waists. They are usually white or brown, and unlike ants, have no "wasp waists."

Another wood pest is the carpenter ant, which to the untrained eye looks like any other ant. They usually show their presence in the house because they tunnel through the wood but do not eat it, and thus leave small piles of sawdust under wood they have attacked. They often remain outdoors to attack power line poles and trees. They are harder to eradicate than termites because the nest must be found and destroyed.

Powder post beetles are characterized by the presence of fine to coarse powder that is pushed to the outside of the wood through tiny holes in the surface, or is packed tightly in the galleries made in the wood.

The old house borer is an eastern United States pest, from Massachusetts to Florida, and also can be found in Mississippi, Louisiana, and Texas. None have been found west of the Mississippi River except in Texas. The term old house borer comes from its infestation of old houses in Europe, but it can attack new construction as well. The larvae of the beetle are the culprits and are hard to find until they have done the damage. They are detectable by a rapping or ticking sound in the wood while they're working; blistering of the wood; powder near the wood surface, found by breaking the surface; boring sawdust on surfaces below infected wood; surface hole of about ¼ inch; and beetles in the house.

That takes care of the insects. Then there are the rot-producing fungi — mold and mildew — which grow in moist areas at temperatures between 70 and 85 degrees. The decay makes the wood spongy, and a poke with an ice pick or other sharp instrument will indicate this. Decay is often called dry rot, which is a misnomer, because fungus needs moisture to grow and cause decay. The term dry rot has grown popular because the damage is often discovered after the wood has dried out and the fungus has died. The decay usually occurs just below the surface of the wood and works its way into the heart of each timber.

One thing about rot-producing fungus: it cannot grow when the wood is dry, and dry, sound wood, even next to rotten wood, will not rot. The most important thing to consider in dealing with fungus is "dry."

Best protection against insects and fungi is preventive, through the use of particular woods, treatment of the wood, sanitation, and design. Fungi will not grow in dry wood (20 percent moisture or less), so make sure the wood you use is sufficiently dry. Kiln-drying temperatures will kill fungi. Do not use wood that is discolored by a bluish or grayish color. This is caused by stain fungi, and while such fungi may not cause decay, their presence shows that conditions exist to encourage the growth of decay fungi as well.

Heartwood is more resistant to decay and insect invasion than sapwood. Heartwood is getting difficult to obtain, for obvious reasons. Species of wood such as redwood, cypress, and cedar are naturally resistant to decay and insect attack, but availability, expense, and strength of the wood may require the use of other woods.

Sanitation is also important. Don't allow any holes on the inside or outside of the foundation to be filled with any kind of wood, and remove all wood scraps before backfilling. Insects can get their start in such buried debris.

Wood should not come in contact with the earth. Some specifications allow treated wood or redwood to be in contact with the soil, but this is not a good practice. Stair carriages should be on concrete piers at least 6 inches above grade.

Wood on wall members should be at least 8 inches above the earth, and for extra safety (and peace of mind) this distance should be 18 inches. Inside the foundation, in crawl spaces, the bottom of joists should be at least 18 inches above the ground, and for extra safety and ventilation purposes, this should be at least 3 feet.

Crawl spaces should have a vapor barrier of 6-mil polyethylene or roll roofing on the earth floor. Lap polyethylene at least 2 to 3 feet and bring the sheets up against the foundation wall at least 6 inches. Secure with weights. Lap roll roofing 6 inches and seal with roofing cement.

Ventilate the crawl spaces with cross-ventilation, as described in Chapter 7. Ventilate attics as described in Chapter 18.

Roofs should have sufficient overhang so that rainwater does not run down the walls. Gutters and downspouts, if used, should guide water away from the house. Foundations must be properly drained. Posts and other wood members standing on concrete should be elevated with runoff slopes.

The basic rules are simple: no wood in contact with the earth (not even a lattice skirt around a porch or plant trellis against the wall if it's standing on the earth) and no places where wood sits on a horizontal surface where water or moisture can be trapped.

Foundations and other concrete materials should be free of cracks or other joints. If there are joints, they should be filled with cement mortar, and openings around pipes, conduits, and other items that must pierce a wall should be sealed with bituminous sealers or caulking.

Slab-on-ground structures are particularly susceptible to termite invasion. If you plan a slab-on-ground construction, make it the monolithic type, with footing, foundation, and slab all in one piece, or at least the foundation and slab in one piece. Reinforcement will prevent cracking.

If your slab rests on top of the foundation, the construction is also suitable. Not suitable is a slab that sits only partially on the foundation or not at all (entirely on the earth), resulting in a vertical crack between earth or near it and the top of the slab. If you have such a crack, seal it with a bituminous sealer or caulking compound.

A termite shield can be installed over the foundation of a crawl space or basement house; and on top of the slab of a slab-on-ground. Such a shield must be installed before the sill is installed, but no termite shield has been developed that is absolutely effective against termites. Properly installed, a shield can be fairly effective.

Metal, such as copper, aluminum, zinc, and other corrosion- and rust-resistant metals, is applied to the top of the foundation in long strips, extending 2 inches beyond the wall on the outside and 4 inches on the inside. The outside 2-inch overhang is bent down at a 45-degree angle. Half of the 4-inch overhang on the inside is bent down at the same angle (on slabs this is impossible, so the metal must end at the edge of the sill or floorplate). The weight of the sill and the house will hold the shield in place. Where anchor bolts stick through, fill space with bituminous sealer or caulking compound.

The shield, in effect, is like a cat-shield around a tree, requiring the termites to build their tunnels out and around the shield, something they are not inclined to do, and if they do it, the mud tunnels are readily detectable.

Wood that is exposed to weather and is set next to concrete can be treated with a wood preservative that is water-resistant and generally toxic to insects and fungi. It has been found that properly treated wood is not attacked by insects and fungi.

Setup windows and doors are — or should be — made of wood that is treated with a wood preservative, applied under pressure for maximum penetration. You can buy pressure-treated lumber for sills and other members that might be open to insect attack or moisture penetration.

If you can't buy the treated wood, you can treat it yourself. Use material containing pentachlorophenol or copper naphthate. Both chemicals are sold under different brand names. Some are clear and can be painted, others are already colored.

Soaking for a minimum of 3 minutes is most effective of the nonpressure techniques, and is particularly important in treating end pieces, where the end grain is exposed. End grain wood absorbs water like crazy, but it also will soak up more preservative than the sides or edges of a piece of lumber. Or you can use a brush, with two or three coats preferred. Let each coat dry before recoating, and brush the wood generously, keeping a container under the lumber to catch runoff. When treating members, it is important to treat all exposed surfaces, and to treat the wood after it is cut and before it is installed. This adds to the time of construction, but not all wood has to be treated.

Poisoning of the ground is effective treatment against termites and other insects. The use of poisons is extremely hazardous and should be done by a professional pest-control expert. Precautions are especially important when you have slab-on-ground construction. Chlordane is generally used for termite and carpenter ant control, a 1 percent solution in water for termites, a 2 percent solution for carpenter ants. In fact, chlordane is one of the few insecticides now allowed for termite and ant control.

Effective treatment uses many gallons of the solution, which is applied under the slab and along the foundation for slab houses, along the foundation and sometimes under the footings with basement or crawl-space houses.

It cannot be stressed too much that the use of insecticides is hazardous and should be handled by experts. Protection against termites usually is guaranteed a number of years. Protection against carpenter ants and other insects is more difficult, and the guarantees are hard to make. If an amateur applies the insecticide himself, and it's not right up to specifications, the work and cost are virtually useless.

For a publication on termite control, write to "Subterranean Termites," Home and Garden Bulletin No. 64, Superintendent of Documents, U.S. Government Printing Office, Washington, D.C. 20402.

Finally, be nice to spiders. They are probably your greatest protection against insects of all kinds. OK, if you find one hovering over your pillow, or in any part of the living section of the house, take it into the basement. They'll eat many insects in the course of their lives. Also be nice to hornets and wasps, at least if they're in areas away from the house and away from where children play. Wasps and spiders are predators, the lions and tigers and wolves and sharks of the insect world.

chapter 26

The Big Cover-Up

Paint and Stain, Inside and Out

If a piece of unpainted wood gets soaked with water, and the water drips out of it, it will appear dark brown or black. This is not dirt, but the pigment in the wood. And it's the pigment in the wood that streaks areas where water has soaked through the wood, painted or not. This is one reason, a little more obscure than other reasons, why wood, whether it's interior or exterior, should be painted, stained, or protected in one way or another.

Now, there are situations where you don't have to stain or paint exterior wood. When you apply white cedar shingles to the walls, they can be left to weather, usually to a nice, even, gray color. Sometimes, when they are not influenced by salt water, as along seacoasts, they might get a little blotchy, with black highlights. But being cedar, they resist decay, and erode, or wear away, at the rate of perhaps ¼ inch every hundred years. The point of white cedar shingles is that they don't need painting or staining, a big plus in maintenance.

Other woods that don't need covering in any way are redwood and cypress. Redwood is very expensive and cypress is not readily available.

So, white cedar does not have to be finished. But people will continue to paint or stain it or even bleach it, and will often do the same with redwood siding. Because of their excellent resistance to weathering, red cedar, cypress, and redwood take paint well, and are resistant to cupping and checking. Which is all you have to worry about.

White cedar is installed green, that is, uncured, so it contains a lot of moisture. And thus it must be installed several months before it can be painted, or even stained. It will take paint and stain well, once cured, and will take on an attractive gray when bleaching oil is applied.

Red cedar should be painted or stained, and cedar clapboards can be left as is to weather or can be stained, bleached, or painted. Redwood can be treated the same way. The various pines, firs, hemlock, spruce and all the hardwoods are more difficult to paint.

Perhaps you are going to put up some kind of cedar siding with a pine trim. What to do with them? Stain or paint?

Stain is the way to go on exterior walls. You can get clear stains, whose color is obtained from dyes (the color selection is limited), and you can get pigmented stains, with virtually no limit to colors, including an off-white. The pigments, while adding color, also obscure the grain of the wood a little, but both pigmented and clear stains penetrate the wood, not only to preserve it (some stains have a wood preservative added) but also to make it water-resistant. And the penetration of the stains into the wood does not leave a film on the surface. When paint is applied, it forms a film, and that is how it protects. Paint wears off, can peel, and cause all kinds of problems, and usually needs to be scraped, sanded, and wire-brushed before repainting, which involves application of a primer and finish coat.

Stain does not require this. Having no film, it wears away, and when restaining is necessary, you simply apply the stain again. It's a great plus in maintenance.

Stain works best on a rough surface. Shingles have a rough surface, and red cedar is ideal for staining, as is resawn and unplaned clapboard or other kinds of rough siding. Stain does not penetrate as well into a smooth surface, however. If you plan to stain clapboards, put them on "inside out," that is, rough side to the weather.

Two coats of stain are needed, and restaining is not necessary for three to five years and more.

Stains are less likely to be affected by moisture penetration than paints, particularly from the inside.

Putting a natural finish such as varnish on exterior walls is not recommended. Even the best marine varnishes tend to turn yellow, and will last two or three years at the most. When they peel, they're likely to expose unprotected wood, which will turn dark as it weathers. Then, before refinishing with varnish, the wood must be sanded to its natural state; otherwise the varnish will turn the weathered wood black.

Paint provides good protection against water penetration and comes in an infinite variety of colors. It is not a preservative; rather it forms a film on the surface to prevent water penetration.

Some experts recommend a water-repellent preservative treatment before painting. This would include an application of pentachlorophenol, then a primer, and one or two finish coats. The primer recommended is a nonporous oil-base material. Then apply one or two coats of a top or finish coat. Apply the finish coat within two weeks of the primer, and if necessary, the second coat within two weeks of the first.

If you complete your house in the fall, and cannot do the entire paint job, at least put on a coat of primer so it will protect the wood. In the spring you can reprime if necessary and then paint.

When using oil paints, do not paint where the sun is warming the siding. Follow the sun around the house, always painting in the shade. With latex paint, don't paint in the coolness of the spring or fall, and avoid painting where moisture occurs before the paint has dried. Of course, when using oil paints, don't paint where there is any moisture either.

There is a running controversy among manufacturers of latex and oil paints, which seems strange since most manufacturers sell both. But the argument is that one is better than the other, for various reasons. For instance, the latex people argue that latex "breathes" and allows water vapor to go through it without injuring or blistering the paint. The oil people maintain that the primer has to be oil anyway, so use oil, which gives a better film and protects better. The latex people retort that a latex primer can be used with latex paint.

The argument is academic. With the proper primer, and a properly constructed and insulated house, either paint will last five to seven years.

But the proper construction is important. Biggest cause of failure of any new paint system is water vapor penetrating the walls from the inside and pushing against the paint film. This is usually the case with houses without insulation and a vapor barrier.

If you have insulated your walls properly, and have installed a full vapor barrier toward the heated part of the house, water vapor will not penetrate the barrier and reach the paint to push through it.

Whatever kind of paint you choose, make sure you use primer and finish coat that are compatible. To do this, make sure they are of the same brand. And when repainting, use the same kind (latex or oil) that you used originally, and if the original paint was of decent quality, stick with the same brand.

It used to be that if you used latex over an old oil-based paint, disaster resulted. Now the latex people are recommending their special primer first if you're going to put latex over oil. Oil over old latex is said to be not as disastrous as the other way around used to be.

When repainting, all loose paint must be scraped and sanded off, and bare wood spot-primed. When repainting with latex, apply a primer overall before applying the top coat.

Setup doors and windows are often preprimed, and can be painted with a coat or two of finish paint. Sometimes the frames are covered with a vinyl shield, which is claimed to be thick enough to last a lifetime. The word "lifetime" is meaningless. Does it mean the life of the occupant? That means the house itself will far outlast the vinyl. Does it mean the life of the house? That could mean one, two, or three hundred years.

Preprimed or vinyl coated window and door frames are generally white, in contrast to or complementing the siding color. If the wood is neither painted nor vinyl-coated, it can be painted a contrasting color, the same color as the siding, or stained to match or contrast.

Here's where the species of the wood gets involved. Outside trim is usually made of pine (it doesn't matter what kind of pine is used) and is a little tougher to paint than other woods, but this is a minor problem. The problem comes when you buy wood for the trim. Clear pine, with no knots, is best to paint, because the knots in unclear pine (No. 1 common and lower grades) will bleed through the paint and show ugly brown circles or disks. But when you price clear pine, you might think twice because it is probably the most expensive of the softwoods, its price on a par with the redwoods. If there isn't a great deal of trim, then go ahead, but there is usually enough to justify a less expensive grade.

But never fear. You can paint the knots

with two coats of a stain killer, fresh shellac, or aluminum paint. Use of a stain-killing cover is no guarantee that the knots won't bleed through, but chances are the problem will be solved.

An important part of finishing wood, exterior or interior, is filling countersunk nailheads. On clapboards, use siding nails which can be countersunk (indented into the wood). On exterior trim, use casing nails, with semismall heads that can be countersunk easily. Glazing compound is an excellent "putty" for filling nailheads. Pigmented exterior stains and paints will cover the compound nicely, and if you use a clear exterior stain, you can use color in oil to mix with the compound so the filled nailheads will not show through the stain.

Then, paint the trim with a primer and one or two coats of trim paint, which is shinier than regular house paint. If the trim is not white, you can add color to the primer to match the color of the top coat, and you may need to give only one top coat.

Trim can also be stained to match or contrast with the siding color. If the stain is an "earth" color like brown, or dark red, you probably won't have to seal the knots with stain killer. If the stain is light-colored or pastel, then the stain killer is necessary.

The only other paint jobs are doors and accessories such as shutters. The same rules for painting apply to them. Make sure you paint the edges of the door (top, sides, and bottom) with oil primer and top coat to seal out moisture. Then paint the inside surface with an oil undercoater and then interior enamel. Otherwise moisture can enter the wood and swell the door. When a door swells it sticks, and when it sticks, it's miserable.

Inside, painting and staining are much simpler, because there is no need to protect against weather, only wear. But even for inside doors, the edges should be painted to keep out that wood-swelling moisture.

Millwork, interior trim such as baseboards, window and door casings, and crown moldings, are made of clear pine and can be painted, finished naturally, or stained and varnished.

Some millwork comes in mahogany, about the only hardwood available for trim. It is modern in style, and is best finished naturally or stained and varnished.

On interior trim, use finishing (headless) nails and countersink the heads. Use wood putty to fill the holes. Wood putty is a powder that can be mixed with water. It does not shrink, and both paint and pigmented stain will cover the putty. If you want to stain with clear dye stain, mix the putty powder with latex-based stain of the color desired instead of water. A little experimenting will bring you to the color to match the stain. Incidentally, holes filled with a material darker than the surrounding stain will not be as conspicuous as holes filled with a material lighter than the surrounding stain.

The putty can be smoothed with a damp sponge or sanded after it is dry. Putties that are thinned with ether and other solvents are good, but they are expensive, and they tend to stain the wood, even if you sand heavily, so when you put on the dye stain, the putty stain will show up.

One warning about filling holes with a putty made with stain: it can color the surrounding wood much darker than the stain you desire. So, before filling the holes, stain the wood to the desired color, then give it one coat of varnish, which seals the wood, and then apply the putty. Any putty you smear on the surface can be wiped off with a damp sponge.

To paint interior woodwork (pine), apply an enamel undercoater first, then one or two coats of semigloss or gloss enamel. Semigloss is the best choice, because it wears well, is easy to clean, and does not glare. One or two coats of enamel can be used, and the undercoater can be tinted to the same color as the enamel, so perhaps only one coat of enamel may be needed.

If the wood has knots, the knots must be sealed with two coats of stain killer, shellac or aluminum paint. Some pines are so resinous that it won't hurt to give all surfaces a coat of shellac. This also seals the surface to give the undercoat and enamel a smooth surface for a perfect paint job.

Interior undercoater and enamel can be oil- or latex-based. Latex paints are not only durable and washable, but also go on easily and require only water for cleanup. Woodwork should not be painted with a flat paint, although some latexes, known as satin or eggshell, are nearly flat, enough so to give an elegant surface but with just enough sheen to increase durability and washability.

Both softwood and hardwood trim can be finished naturally, with high or semigloss varnish, or stained and varnished.

Stains come clear; that is, the color is provided by dyes that allow the grain to show through. The best dye stains are penetrating oil stains, which will not raise the grain of the wood as water stains will. Other types of stains contain pigments for color, which obscure the grain somewhat, and come in considerably more colors than the dye stains.

Any type of stain can be applied with brush or cloth and wiped off to the desired tone. The fast-dying stains with a latex base must be wiped off much more quickly than the oil stains because if they aren't they'll look almost like paint.

To make sure the penetrating stains go into the wood evenly, wood should be sealed with a penetrating sealer first. A coat of shellac thinned half and half with denatured alcohol is a good sealer. Do not use varnish stain to try to do the job of staining and varnishing at the same time. The result will be very poor.

Urethane varnish is among the best of the clear varnishes. Generally, the first coat acts as a sealer while the second coat finishes the surface. Some brands require thinning with paint thinner or mineral spirits for the first coat. Sometimes three coats are necessary for a smooth, even surface, particularly with kitchen cabinets, where durability and washability are necessary.

Semigloss urethane varnish is slightly less durable than high gloss, but the difference is not enough to justify using high gloss when you want a good surface in the kitchen. Semigloss varnishes, however, are slightly cloudy, which may be objectionable. Varnishes also come in flat finishes.

Ceilings can be painted with a flat white latex ceiling paint, using a brush to get into corners and edges and a roller to do large areas. Two coats are usually needed. You can also buy texture paint, a thick, inexpensive latex paint that can be brushed to look like a swirled plaster finish, or stippled, or worked into any design your artistic ability can produce.

Walls take flat latex wall paint. The first coat primes and seals either plaster or plasterboard, the second coat finishes. Semigloss paints are usually applied to kitchen and bathroom walls for durability and washability. Oil paints can be used, either flat or semigloss, but they need a special primer before the finish coat is applied. Plaster walls should cure for six months before they are finished.

Painting hardwood trim (it's interior, never exterior) is not recommended because of the cost of hardwood and its attractiveness when stained and varnished or varnished naturally. Besides, many hardwoods have open pores that show up when painted, making the wood look as if it had the measles. Hardwoods with open pores include ash, chestnut, elm, hickory, mahogany, African mahogany, oak, and walnut. Close-pored hardwoods include cherry, gum, maple, sycamore, and birch.

If for any reason you are going to paint open-pored hardwood, you must fill it with a paste wood filler for a smooth surface. The filler is applied with a brush across the grain, then with the grain. Immediately after the filler has lost its gloss, usually in a few minutes, wipe off excess filler with burlap or similar rough cloth, first across the grain to pack in the filler, then lightly with the grain. If the excess filler dries, it must be sanded off.

If you finish open-pored hardwoods naturally or with a stain and varnish, it's a matter of taste whether to fill the pores. An open-poured hardwood given several coats of varnish will not have the same dead-smooth, even sheen that a filled wood or close-pored hardwood or softwood has. But some experts maintain that this is a part of style and decorating. So take your pick.

Floors are usually varnished or sealed and waxed. They can be stained first, or left natural.

They must be sanded first, to achieve as smooth a surface as possible. Although oak is an open-pored hardwood, do not fill the pores with paste filler. The resulting finish is not quite as silky smooth as it would be if the pores were filled, but the surface will not show scratches and other marks as readily as it would had it been filled.

For a varnished surface, which gives a very hard, long-wearing surface, apply three coats of high-gloss urethane varnish. Here is where the less durable semigloss should not be used. Varnish wears well, but it does scratch under hard use, and it is difficult to spot-refinish in heavy-wear areas.

So, when heavy traffic areas are worn, a varnished floor must be sanded thoroughly (by hand or small portable sander, not the floor monsters used on a new floor) and one or two new coats given. This will renew the floor, but it probably can't be done this way more than once. When it really gets bad, the whole surface must be removed and new wood exposed and smoothed, with floor sander and edger.

When using urethane varnish, follow label instructions. When they say to give the second coat not less than four hours after the first but not more than twenty-four, they mean it. If the first one or two coats gets too dry (hard), any following coats will peel off like onion skin, unless the previous coat is hand-sanded to remove or reduce the gloss.

Some experts prefer a floor finish of sealer and wax, achieved by mopping a penetrating sealer liberally on the floor, letting it sit for fifteen minutes or so, then wiping up the excess. Wiping the excess is important; otherwise the surface will stay sticky. Anywhere from three to six or even ten coats of sealer are recommended, each applied in the same manner; the more coats, the glossier the finish. Sealer must be protected with a wax coating. Use a paste wax and buff with a buffing machine to harden and shine the wax. A sealed and waxed floor penetrates the wood and will not scratch. The wax does get dirty, however, and must be removed with turpentine or wax stripper and a new coating applied every six months or so.

Similar surfaces can be put on other wood floors, such as fir or pine. Fir flooring is not too attractive, in some opinions, but it will stand up well. Hard pine is durable, but is not, in some opinions, very attractive unless it is stained dark and then finished. Other types of pine are attractive, especially when stained. To simulate the old Colonial pumpkin (color) pine floors, a stain must be applied.

Outside floor surfaces, as on porches, should be treated with a wood preservative and either stained or painted. Some of the wood preservatives, with copper naphthate bases, often contain their own stains and can be attractive. However, such stains are flat, with no gloss, and tend to show footprints.

Finally, everything is done inside, except you may want wallpaper. OK, fine. If the walls are plaster, all you have to do is apply glue size, a powder you mix with water, and then apply the wallpaper. Glue size seals the wall, and it does some funny things: it makes the wallpaper slip better when you are applying it so you can slide it into position better; it makes the paper come off easier when you want it to.

But plastcrboard is another story. Plasterboard has its own paper surface, so it must be sealed with more than glue size. It must be waterproofed with a flat, interior oil paint. Use white; it is a good neutral background for wallpaper. The paint will help cover any marks made on the plasterboard, which could show through a light-colored, lightly patterned paper.

Then size the walls with the glue size, and put on the paper. The waterproofing will allow you to remove the paper without ripping off the plasterboard paper and ruining the wall.

Wallpaper itself comes in so many kinds and styles that it will bewilder the beginner, but sooner or later you'll hit on what you want. Some years ago, paper was considered very middle class; elegant houses were painted on the inside. Not any more. Today, wall coverings are considered "in," and very elegant, or bright, or mod, or crazy. But one thing these coverings have in common: high price.

Not too many years ago, wallpaper was just that: printed paper that was put on with

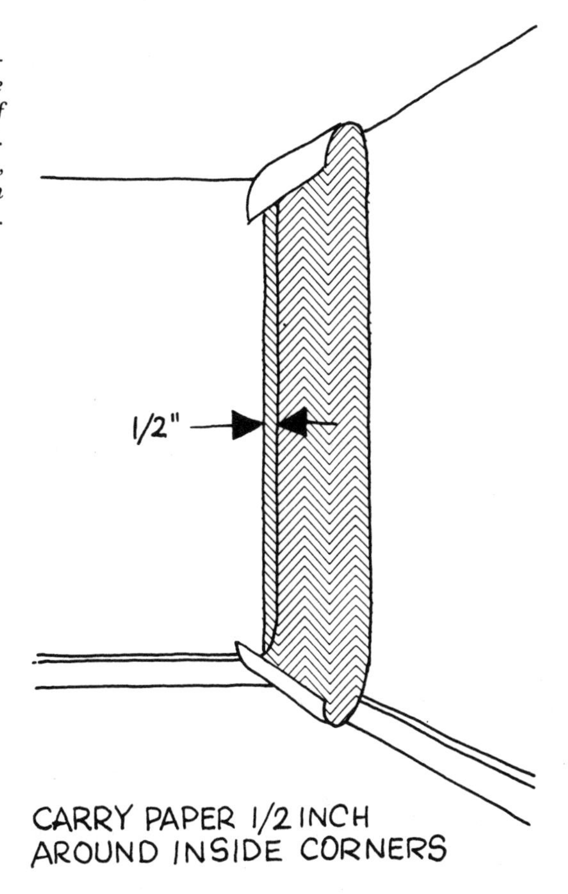

FIGURE 59. When hanging wallpaper, start at a corner, with ½ inch of the paper strip overlapping the corner. Make sure the far edge of the first strip of wallpaper is plumb (vertical) by using a spirit level. The vertical line must be determined for each wall, and when the first strip goes on the next wall, it can overlap the ½-inch section, if necessary, to be vertical.

wallpaper paste. No more. Yes, there are still paper wallpapers, and they are relatively inexpensive, but there are also vinyl-coated paper, canvas types, flocked paper, and foil-backed paper. There is even strippable paper, which can be removed by simply grabbing a corner and pulling.

Probably the best development with wallpapers is the prepasted paper, which needs only to be soaked in water (long, narrow trays are sold or provided for this purpose) and hung. Prepasted papers are very good, and the majority of papers sold today are prepasted. You can't get all patterns and types prepasted, so if you buy a paper that isn't prepasted, make sure you buy the paste designed for the paper. Some of the heavier papers need a heavier, stronger paste than others.

Wallpaper comes in double and triple rolls.

A single roll covers about 32 square feet, and sometimes the double or triple roll is not all in one piece. When you run into a cut roll, you're not getting gypped, because extra paper is provided.

Wallpaper must be hung plumb. If the ceiling, floor, or another wall is not level or plumb, ignore it. Cut the paper several inches longer than the distance from ceiling to baseboard. Cut the paper where you want the pattern to disappear into the ceiling. A utility knife cuts paper very well, but to cut wet paper, you'll need a serrated circular knife, something like a pizza cutter, which makes perforated lines that can be torn, with care. With the canvas-type papers, a utility knife is necessary. Replace the blade whenever it tears paper or pulls when cutting. Other tools you'll need are a paste brush, bucket, and a smoothing brush.

To hang the first strip, start in a corner (Figure 59), and mark a plumb line on the wall, away from the corner, ½ inch less than the width of the paper strip. Use a plumb line or long level to make the mark plumb. For instance, if the strip is 21 inches wide, make the mark 20½ inches from the corner. This will allow half an inch of paper to turn the corner. Any more paper turning the corner will result in wrinkling and a frustrating mess. You'll also need a wallpaper table. You can make one by taking a door off its hinges and putting it on the backs of two chairs or any other support for a comfortable working height. To prevent paste from getting on the table, cover it with newspapers. Place the paper facedown and apply paste at the bottom half first (the back of the paper has arrows to indicate which direction it is supposed to go in) by pasting down the center of the strip, lengthwise, and then paste the edges by brushing from the center to the edge in both directions. Then fold the bottom half over on itself. This is done because most tables (you can even use a kitchen or dining room table, with plenty of newspapers) are not as long as the length of the strip. Then paste the top half.

When applying the paper, line it up with the plumb mark. Let the top overlap on the ceiling. Smooth the paper with the smoothing brush, working out bubbles and wrinkles by pushing them toward the edge of the paper. When you're partway down, and the paper will not move, unfold the bottom half and continue. Don't worry about tiny wrinkles; they'll disappear when the paper dries. When the paper is all smoothed off, cut it at the ceiling and baseboard.

A trick with cutting is to put the edge of a wide putty knife (6 to 10 inches wide, which makes it a smoothing knife) into the corner where wall and ceiling meet. Then, run the cutter or knife along the edge of the smoothing knife and you will get an accurate cut.

After you've smoothed and cut the paper, go over it with a damp sponge to remove excess glue from paper and along the ceiling and baseboard. Keep a bucket of water for the sponge and change the water often.

Sometimes the paper will have an extra long "drop" in its pattern, which means you may waste a lot of paper when lining up the pattern of the next strip to match the pattern of the first strip. To avoid this, instead of taking the next strip from the same roll, take it from a separate roll. You'll be amazed how much paper you will save.

When you match a pattern, match it at the top, or at least eye level; it may be a tiny bit off at the bottom, but it won't be noticeable. It is something that cannot be avoided. If the pattern doesn't seem to match no matter what you do, take the paper back for replacement.

When papering around door and window frames, you can measure and cut the strip to fit the opening along the frame. Then use the part you cut off to fit over and under the window frame. You can do the same thing when you come to another corner. For instance, if the space you need to fill a corner is only 10 inches wide, cut a strip 10½ inches wide so it will turn the corner by ½ inch. Then use the piece you cut off to continue on the new wall. When continuing on a new wall, you must draw another plumb line, because if the walls are not perpendicular to each other, the continuation of the strip of paper might not be plumb.

Around frames, you can rough-cut the paper for the opening, so you won't have so much paper to handle; then cut around the frame when the paper is on the wall, just as you did at the ceiling and baseboard.

With prepasted paper, follow instructions: let the strip soak for a moment or so in the pan, so that all surfaces are covered and the prepaste is wetted, then hang it in the regular manner.

With paper without a matching pattern, you don't have to worry about matching it, naturally. If the paper is tweedy, then you can patch and use bits of paper here and there. If the paper is completely plain, the less patching you do the better because the seams tend to show with no pattern at all.

And remember, you're working 12 to 18 inches away from the paper when you're installing it. Once it's up, and you're sitting in your easy chair, you really won't notice errors, if there are any. You may know where you goofed, but you can learn to be philosophical about it.

If you get a wrinkle that nothing seems to eliminate, cut or tear the paper along the wrinkle to smooth it out. A torn edge is less noticeable than a cut one.

And if the cut or tear is high on the wall, lap the bottom edge over the top one, so as you look up you won't see the joint. If the cut is toward the bottom of the strip, lap the top edge over the bottom edge of the cut. There are lots of ways to fool the eye.

Wallpapering is exacting, somewhat fussy work, and grown men have been known to cry over the frustrations. But the nice thing about it is that when you've put up each strip, you don't have to do anything else to it. Like prefinished paneling, it is finished.

And it also means that you're nearly home with the old house. Except for a few odds and ends, it is already your castle.

chapter 27

Odds and Ends

Steps, Porches, Garage, Driveways, and Walks

Your house is finished, and probably could have been occupied, with that coveted occupancy permit, even before you put in the woodwork and painted, wallpapered, and finished up in general.

But though it is finished, it really isn't done. There are such things left to do as outside stairways, porches, the basement floor, garage or carport, other outbuildings such as garden and toolhouses, and sidewalks and driveways. Not to mention landscaping, which is grading of the property, sodding or sowing grass seed, and installing trees and shrubbery. Most of these things can be done at your leisure, or when you can afford them, or both.

Many houses have no porches at all, but rather just a "stoop," roofed or not, at front and back entrances. Whether you build a porch or not is a matter of taste and the house's design. The days of the front veranda, a long, sweeping, covered porch made of wood, complete with wicker furniture, are gone forever, unfortunately, although many houses are still standing with such attachments. They should stay, but they are expensive to build, and to maintain.

So, let's tackle the front, first. Generally, the nearer the door is to the ground, the simpler steps and stoop are to make. Many front entrances are just a slab with a step or two.

A front slab made of concrete, on the ground or several steps high, should be built with footings and foundation, just as the main house is. Sometimes such a foundation is included, sometimes it is added. If it is added, the footings must go below the frost line, with a regular foundation, and slab poured on 6 inches of gravel on unexcavated ground. Never pour directly on filled earth, unless it has been tamped thoroughly. Otherwise the slab could collapse as the ground settles. Steel reinforcing rods or mesh can prevent this. With footings and foundation, a slab can be tied in with the main foundation of the house, using steel rods. Also with proper footings and foundation, the slab can be separate from the main house.

The concrete steps should be built as a part of the slab, and have their own footings and foundation. Steps should follow the step formula; that is, two risers and one tread should total 25 inches. Forms can be built for forming steps and slab. The slab should be at least 4 inches thick, and if the slab is several steps high, rubble (not including wood) and other fill can be put in the hole inside the foundation to conserve concrete. The steps should be solid, without too much rubble thrown in the forms. Steps should pitch away from the house about ⅜ inch. The concrete should be finished with a wood float for a rough, nonslip finish.

The slab itself should be big enough to accommodate an out-swinging screen or storm door, so should be a minimum of 4 x 4 feet. Steps should match the width of the slab. If the "stoop" is going to be covered with a roof supported by columns, it should be bigger to be in proper proportion to the house itself.

You can buy ready-made concrete steps and stoop, but they must be set on a foundation.

The slab should be either as level with the door threshold as possible or at least one full riser lower than the threshold. An "in-between" riser height from slab to threshold is an unnatural transition and will promote stumbling, sober or not.

A slab and steps of brick provide a good decorative effect. The slab should be lowered to accommodate the thickness of a brick, about 2 inches. Bricks can be laid in fresh concrete by embedding them slightly. The concrete will hold well, and when it has set, the joints are filled with mortar.

There are several patterns of brick paving, which also applies to sidewalks or patios. The pattern can be a common bond, with the short joints of the bricks staggered, or a stacked bond, with the bricks lined up like soldiers in rank. A herringbone or basketweave layout is also decorative. When a brick is at the edge of a slab, it should be placed to form a border, with the ends of each brick overlapping the edge of the slab by ½ to 1 inch, as a drip edge. At corners, a patio block or stone should be cut to fill the square to prevent corner bricks from falling out. It's a good idea to set the edge bricks on their narrow edge. This would require the center part of the slab to have enough mortar or sand concrete to allow the tops of bricks on their wide edge to be level with the tops of the bricks on their narrow edge.

With steps, a good combination is a 7-inch riser and a 10½-inch tread: the base of the first riser is a row of bricks on their wide side, set in a ½-inch mortar joint so that they are parallel with the length of the step. They are set so they form a tread of 10 inches. Then, the cavity behind the row is filled with concrete.

The tread itself is made up of a row of bricks on their narrow side and perpendicular to the length of the step. The location is such that their ends extend 2½ inches from the edge of the next riser. Then, a second row of bricks is laid on their narrow edge, also perpendicular to the length of the step, to form the rest of the tread, and overlapping the supporting row by ¾ inch. Thus, with a ½-inch mortar joint, the 2½ inches of the inner part of the step plus the 7½ inches of the full brick totals 10½ inches. Brickwork is tricky to do, but the appearance is worth it. It may be worth your while to hire a mason and let him do it, with you acting as mortar mixer and hod carrier.

When laying the steps, the mortar should be "wet," at least wet enough so you can butter the edges of the bricks you will lay, but not so wet that the mortar gives too much when you put the bricks on top of it. But when filling the joints of bricks laid on a concrete floor, the mortar should be almost dry, just about crumbly. Then, when you pack the mortar into the joints, it won't get all over the face of the brick. Pack mortar thoroughly and completely until it is flush with the face of the brick. Then, when the mortar is set for 10 minutes or so, with a pointing tool, an S-shaped flat steel rod, strike the joints by dragging the tool over the mortar, creating a concave joint. The more you work mortar the wetter it becomes, so keep the working to a minimum.

Mortar that gets on the face of brick can be removed by treating with muriatic acid. Dilute it at least 2 parts of water to 1 of acid, and pour generously over the face of the brick. It will fizz up on the mortar and when it stops fizzing you can scrub it with a broom. Then wash it all off. If any mortar remains on the face of the brick, repeat the process.

Used or new brick is equally good for outdoor purposes, but make sure you buy a water-struck brick or at least one that will stand up to the weather. The old-fashioned common brick, when laid on a horizontal surface, will not stand up to weather, even though it will last a long time in a vertical wall.

Sometimes a stoop and steps are made of concrete and topped with a veneer of brick. The technique of application is the same, and whether you put the brick on its wide or narrow side depends on the height that you made the concrete riser. A step is best when the brick is laid on its narrow edge.

Roofs for porches can be flat or gabled. They are attached to the house when the framing goes up, or at least when sheathing is in place. Flashing must be applied to the roof where it meets the wall. Usually, columns or pillars support the two outside corners. Columns complete with base and capital come ready-made, in painted aluminum. You can make a simple column with a 4 x 4 or even 6 x 6 post or make your own box column by nailing nominal 1-inch boards in a long box square. Make sure the box is square. You can taper each board slightly for a more graceful appearance. The Greeks actually made their columns bulge in the middle because the bulge fooled the eye into believing the column was properly tapered.

To complete a roof, there should be some finish for the inside corners, too, where the roof meets the wall. This is accomplished with pilasters, half-pillars attached to the wall and supporting the roof. All pillars can have a simple base and capital made of square nominal 2-inch lumber 1½ inches larger on all four sides and finished with a ¾-inch quarter round between the pillar and underside of the capital.

Where pillars touch concrete, they should be treated with wood preservative and flashed to prevent water penetration. They also can be raised slightly for drainage. If a 4 x 4 or 6 x 6 post is used, it is usually fitted over a pin or anchor bolt secured in the concrete. One or two galvanized washers are put on the bolt to keep the post off the concrete.

Back porches these days are sometimes more elaborate than front stoops. They can be built when the house is or added later. If a concrete floor is used, it should be at least 3½ inches thick and with a proper foundation and footings. Wire mesh or steel reinforcing rods should be installed in the concrete. The slab is poured on 6 inches of gravel.

Roof and posts are installed the same way as in the front, and roof and walls, if any, are installed as described in the chapters on wall and roof framing.

Porches can be left without railings, which can be installed later, or walled in and fitted with regular windows or screened.

A railing consists of two 2 x 4 or milled rails, spanning the space between the posts, with the bottom rail set on blocks to keep it off the concrete and to provide drainage. The rails are separated by balusters, vertical spindles made of 2 x 2s or regular milled balusters 1¼ x 1¼ inches. In combination, the railing is called a balustrade. The height of the balustrade can be 24 or 30 inches, depending on the scale of the porch.

A flat-roofed porch is sometimes a little too plain for the front of the house, so you can jazz it up with a balustrade around the roof's edge. Such a balustrade can be shorter than the floor balustrade, or at least in proportion to the roof and the size of the porch. Corner posts are 4 x 4s, and should be integrated into the frame of the roof, or at least installed before roofing shingles or roll roofing. Posts must be flashed and treated with roofing cement to prevent leaks.

Sometimes when a balustrade is fairly long, a row of balusters parading along like soldiers is a little dull. So, interrupt them in a symmetrical way by installing an X in the center of the run, made of balusters. You can make variations of the X according to your whim, or the length of the balustrade. One simple design is to superimpose a cross over the X, a design known as a star in the building trade. It also looks like the British Union Jack.

The basement floor is usually not poured until most of the house work is done. It should be laid on 6 to 18 inches of gravel, particularly if there is a high water table or poor drainage in the ground. This much gravel will help prevent water from rising against the slab, because when that happens, the pressure is irresistible.

Also, lay a 6-mil sheet of polyethylene over the gravel and under the concrete, which should be at least 3½ inches thick, with reinforcing rods or steel wire mesh (6 x 6–inch holes). The polyethylene sheeting will prevent water from percolating up through the slab. The floor finish should be smooth; that is, finished with a steel trowel.

The floor should slope toward drains, if any, that should lead to storm sewers. If there are no storm sewers, the drains should lead to the foundation drainage system. If the ground is especially wet, or you expect it to be, build a sump in the cellar floor. This is a hole 18 inches square and 18 or so inches deep, with concrete sides and a gravel bottom. The floor should pitch slightly toward the sump. Then, if flooding occurs, the water will go into the sump, where it can be pumped away by an automatic sump pump. The water can be pumped through plastic pipe at least 10 feet away from the house or into the drainage system, storm sewer, or into a dry well, which also must be at least 10 feet from the house. Never drain flood or rainwater into a sanitary sewer system or a septic tank.

Another value of the sump is that if the water table, or level of underground water, gets near the flood slab, the sump acts as a safety valve, filling and being pumped away before it can exert a concrete-breaking pressure against the slab.

Garages are common in cold areas; carports are popular in warmer areas. With today's modern cars, designed to resist a lot of cold weather, plus people's habits of replacing their cars before they reach maturity, garages are less important than in the past.

Building codes sometimes restrict the location and design of a garage. Some houses are designed with the garage as a part of the house, either on or below grade, the advantage being that the garage is warmer in cold weather and it is convenient and safe to go from the garage directly into the house. The disadvantage of an attached garage is that auto fumes may leak into the house. The auto user should, in such a case, move his car out of the garage immediately after starting it; no warm-ups in the garage. And when he drives into the garage, he should cut his engine immediately on stopping. If these safety rules are followed, there is no danger of asphyxiating the family in the house, but still, there can be annoying odors.

Some building codes require a fire wall between garage and house, such as concrete block or other fire-resistant material.

Detached garages can be completely independent of the house, or they can be "semi-detached"; connected by a breezeway or covered walk to protect the driver making his mad dash to the house during a storm. A detached garage that comes close to the neighboring property may have to have a fire-resistant material for the wall facing that property.

Size is important for a garage. Although the days of the monster auto may be numbered (a good thing), the minimum depth of a garage should be 22 feet (outside dimensions). Minimum width should be 14 feet for a single car garage and 22 feet for a double garage to prevent the driver or passenger from trying to squeeze through a car door that can be opened only a few inches.

When a shop or storage space is planned, the garage should be planned big enough to accommodate it. The space can be at the side or far end of the garage.

Popular doors for garages are sectional, designed to rise overhead on tracks, manually or electrically operated. They come in standard widths and heights; tracks are adaptable for garages that have a low overhead space.

Construction of a garage is the same for a house on a slab, using 2 x 4 studs and 2 x 6 or 2 x 8 rafters, depending on span and pitch of the roof. Exterior finish can match the house, and the interior can be left unfinished. When the garage is a part of the house, the ceiling and walls that are common with the house must be insulated. And because insulation will not stand up to wear and tear, the wall should be finished to protect it.

The slab should be put on a regular foundation with footings below the frost line, and should be at least 4 inches thick, with steel mesh or rod reinforcing. It should pitch about 2 inches from back to front, or about 1 inch in 10 feet, to allow drainage of rainwater and snow melting off the car. The edge of the slab at the door should be at least 1 inch above the driveway level, with a ramp to prevent bumping over the ridge.

A carport is usually attached to the house, and is only a roof, with walls open. Sometimes the side or end is enclosed for storage space. The construction is the same for a regular garage, except the corner posts should be 4 x 6s, and sometimes, with an extra long and wide carport, the roof is supported by intermediate posts, set over pins set in the concrete floor and kept ¼ inch above the floor to prevent rotting.

Posts and other wood members on or near the concrete should be treated with wood preservative.

Locate the garage or carport to keep driveways as short as practicable. The days of broad front lawns are pretty well done for with the current emphasis on backyard living. So there is no need for elaborate driveways that not only are expensive but also will need plenty of work to clear of snow, if you get much of that stuff in the winter.

Anyway, driveways are designed to keep a car from tearing up sod and earth. They can be concrete or bituminous concrete, also called blacktop or hardtop or hot top. A driveway should be crowned, raised slightly in the center, so rainwater and melting snow will drain off on the sides.

Any driveway should be at least 8 feet wide, and 10 feet is better, particularly if it also is used as a walkway. A double-width driveway should be 18 to 20 feet wide. Sometimes, when there is a double garage, the single driveway can widen to a double width for 12 to 16 feet in front of the garage, so a car can get around one already parked on one side of the superwide driveway. This also applies when the driveway slopes up or down. Twelve to 16 feet of the driveway, where the car sits in front of the garage or carport, should be level for safety's sake.

Concrete should be at least 5 inches thick to stand up under the wear and tear of moving and sitting cars. Use 2 x 6 boards for forms, and to strike off the surface. A wood float should be used to finish the surface off rough enough to be non-skid. Steel 6 x 6–inch mesh reinforcing rods should be used in the concrete, which also should be laid on at least 6 inches of gravel. Blacktop also should be laid on gravel.

Some special considerations are turnarounds and the widening of the driveway at the street, so that sloppy drivers coming home won't ruin that expensive grass.

Concrete should have expansion joints every 12 feet, as well as where the driveway meets the public sidewalk, where the driveway meets the garage or carport floor, and every 12 feet or so in a long run. The expansion joint can be made by leaving a nominal 1-inch board between sections of concrete and then removing it. Make sure the board is well greased so it can be removed. Or you can carve a joint in fresh concrete, a slightly trickier method. The joint can be filled with asphalt-impregnated fiberboard, earth, sand, or any other filler.

The slope of a driveway is dictated by the contours of the land, but the slope should not be toward the garage or carport or house. Always make the slope to the street. A basement garage that is below the grade of the street level is not recommended. In fact, it would be disastrous.

Driveways are expensive. A temporary driveway can be made of crushed stone. Use crushed stone instead of gravel; its pieces have sharp edges, sharper than regular, smooth-sided gravel stones, and such crushed stone will not give as much when driven or stepped on. A crushed stone or gravel driveway is attractive but ultimately impractical. The stones will tend to migrate, one way or another, onto the grass, and it will raise hell with a lawn mower. But if the gravel or crushed stone is left for a year or so, or when you can afford it, all you have to do is remove a little and lay concrete or blacktop on top of it.

Which brings us to the final job, other than landscaping: sidewalks. They can be concrete, blacktop, or something sexy like bricks or flagstone.

They should be at least 3 feet wide, but 5 feet is infinitely better. Three feet will seem like a tightrope. A sidewalk can go directly from the public sidewalk, if any, to the entrance, or can meander all over the place. Or it can go from the driveway to the entrance. And a sidewalk can curve or go straight.

Sidewalk concrete should be at least 3½ inches thick and reinforced with 6 x 6–inch mesh or reinforcing rods. Extra-long sidewalks should have expansion joints every 4 or 5 feet, and should be crowned to shed water. Concrete and blacktop should go over a 6-inch base of gravel.

Or you can get fancy. A popular sidewalk is bricks or flagstone laid in 4 to 6 inches of sand. When laying bricks in sand, you can use most types of brick design, laying the bricks on their wide side and butting them as closely together as possible. Use bricks that will resist the weather; ordinary common bricks of soft clay will break up quickly on an unprotected horizontal surface. Fill the joints with dry sand, and when the sand settles in the cracks, fill them up again. Eventually the joints will be nearly as hard as mortar. Along the border of a brick sidewalk, put in "soldier" bricks, laid their long end down, so they will keep the regular paving bricks from spreading.

Flagstone or patio blocks in sand are not satisfactory, because they will tend to rock and spread.

Of course, both brick and flagstone are more satisfactorily laid in fresh concrete, and their regular (not tightly butted) joints filled with semidry mortar, just as you did on the porch floor.

Bricks in sand, which can be seen in many old cities, particularly the "old" quarters preserving the past, are very serviceable but will tend to heave with the frost in the winter. And every five years or so they'll have to be picked up and relaid.

A sidewalk of more than a 5 percent slope (5 feet every 100 feet) should be stepped. You can make a level walk and then go up the slope with a series of steps, applying the 25-inch formula for two risers plus one tread. Or you can step the entire walk, making platforms several feet long instead of regular treads, with risers about 6 inches high.

With the stepped walkway risen to the heights you want, congratulations! You've built your dream house.

appendix 1

Ya Gotta Know the Language!

Glossary of Building and Other Helpful Terms

In this book and other books on building and home maintenance and repair, and in plans and specifications on houses, you're likely to run into the darnedest terms, so instead of rushing around trying to find out what they mean, find them in this glossary.

It helps to know the language, whether you do most of the work yourself or subcontract it.

Read the glossary first, before plunging into details of construction. While you won't necessarily memorize the terms, you will become familiar with them, and recognize them when reading and when doing.

Abbreviations: A.C., alternating current; d., penny, old English measurement for nails, still used, original was price per 100 nails; D.C., direct current; O.C., on center; O.D., outside diameter or dimension; O.O., outside to outside; R.O., rough opening; X, used for marking lumber where other pieces are to be placed, and also used for "by" as in 2 x 4.

Adhesive: Any chemical or organic material used to fasten pieces together. Types include white, hot, contact cement, panel adhesive, ceramic tile adhesive, resilient tile adhesive, mastic, and glue.

Airway: Space between roof insulation and roof boards to allow passage of air.

Alligatoring: Cracks in a paint surface exposing the coat underneath. *See* checking.

Aluminum foil: Backing for insulation, used as a vapor barrier.

Anchor bolts: Steel bolts embedded in concrete foundation and sticking through holes in sill, held down by nuts; holds down sill.

Apron: Trim immediately below stool, also called stool cap. *See* stool.

Architect: One who designs buildings, draws plans.

Asphalt: Residue of petroleum evaporation. Used for bituminous concrete, roof shingles and, when impregnated with paper or felt, sometimes used as a backing for roof shingles and exterior siding and under finish floors.

Backfill: Refilling earth in an excavated area, usually around the outside of a foundation.

Balusters: Vertical, often turned but sometimes 1¼ x 1¼-inch, members connecting top and bottom rails of a railing system.

Balustrade: A railing on a porch, roof, or stairway, consisting of top rail, balusters, and sometimes a bottom rail.

Band: A decorative piece of millwork used as molding.

Baseboard: Board or milled piece nailed onto wall at floor line. Also called base and mop board.

Basement: Area below grade, usually below first floor of a building.

Base molding: A band that goes on top of a baseboard as decoration.

Base shoe: Quarter round or other type of molding nailed to baseboard where baseboard and floor meet.

Batten: Narrow wood strip to cover joints in vertical boards.

Batter board: A rough board, set in pairs, nailed to stakes to determine the corners and levelness of an excavation. Used in conjunction with surveyor's string.

Beam: Heavy horizontal timber or sill supporting floor joists.

Bearing wall: Any wall or partition that supports any load in addition to its own weight.

Bed molding: Molding at the angle between a vertical and horizontal surface, as between eave and exterior wall. Sometimes a simple board called cornice trim.

Bird's mouth: The notch in a rafter to allow it to sit fully on the top plate of a wall.

Blind-nailing: Nailing through wood so that the nailhead will not show, such as through the tongue of a tongued and grooved board.

Board and batten: Vertical siding in which joints of boards are covered with battens.

Board foot: Measurement of lumber: a piece of wood nominally 1 inch thick, 12 inches long, and 12 inches wide. A 1 x 12 one foot long contains one board foot. A 2 x 12 one foot long contains two board feet.

Boards: Lumber, usually nominally 1 inch in thickness, dressed to nominal widths of 2, 4 (sometimes 5), 6, 8, 10, and 12 inches.

Boston ridge: A technique of overlaying and blind-nailing shingles on the ridge of a roof.

Brace: A board set at an angle to stiffen a stud wall. Usually let in; that is, set in notches in the studs so the surface of both studs and brace are the same.

Brick: Clay blocks, fired for hardness and color, used in building walls, fireplaces, exterior steps, and floors. Water-struck brick is made so that it resists weather. Most finish bricks are water-struck. Common brick is made of plain fired clay and many kinds are not resistant to the weather when placed horizontally.

Bridging: Wood or metal members set between floor and ceiling joists midway in their span. Cross-bridging is a term for members installed in the form of an X; solid bridging is the term for nominal 2-inch members the same depth of the joists themselves and nailed at right angles to the joists.

Butt joint: Where two wood members butt together, end to end or at right angles. The edges of each member are square.

Cap: Anything that tops another member. For instance, the top portion of capital of a column, or the top pieces of molding.

Casement windows: Windows hinged on one side and opening in or out.

Casing: Trim for a door or window, nailed to the jamb or wall for a finish.

Cats: 2 x 4s nailed horizontally between studs to act as nailers for wood-board paneling.

Caulking: Pliable material, dispensed from a caulking gun containing a cartridge, to seal seams, joints, and cracks, for weatherproofing and waterproofing.

Ceiling: Any overhead surface.

Cement: Portland cement, a baked, powdered stone used in making concrete.

Cesspool: A hole in the ground, sometimes lined with stone, with house plumbing attached, to act as a septic tank to break down sewage into sludge (solids), which settles to the bottom of the hole, and liquids, which seep into the ground. Outlawed in most communities. *See* Septic Tank.

Chalk line: A tool, with string saturated with dry chalk, used to mark lines from one point to another on surfaces.

Checking: Cracks in paint. *See* Alligatoring.

Checkrails: In double-hung windows, the bottom of the top window sash and top of the bottom sash, beveled and constructed for an airtight joint where they meet.

Chimney: Vertical tube, usually made of brick, containing a flue for passage of smoke and gases from a fire.

Clapboards: Beveled boards overlapping horizontally, sometimes shiplapped horizontally, used as siding.

Collar beam: Nominal 1- or 2-inch boards connecting opposite roof rafters. Usually spaced every third or fourth rafter, and used to strengthen the rafter system. When collar beams are used for the ceiling over an area under a roof, they are called ceiling joists.

Column: A perpendicular supporting member; also called post or pillar; when made of concrete and a large size, often called a pier.

Combination doors and windows: Usually aluminum, serving as both storm sash and screens. Usually self-storing.

Concrete. A mixture of Portland cement, sand, and aggregate (gravel or crushed stone). Reinforced concrete is reinforced with steel mesh or rods.

Conduit: A tube designed to carry electrical wires.

Cope, or coping: Method of forming end of molding to follow face of adjacent molding in an inside corner, done instead of mitering if the corner is not 90 degrees. Also called scribing.

Corner: Where two perpendicular walls meet. An inside corner has a 90-degree angle; an outside corner has a 270-degree angle.

Corner bead: Metal strip fitting on an outside corner, then plastered. In dry-wall construction, nailed over plasterboard and finished with joint compound. A bead also can be made of wood to protect and decorate the outside corner.

Corner board: Used in pairs, at the outside corner of an exterior wall, against which the ends of siding are butted.

Corner brace: See Brace.

Cornice: Boxed structure at the eave line of a roof, consisting of a facia (face of eave) and soffit (horizontal member of the eave).

Cornice return: The part of a cornice that returns at the gable end of a roof's eave line, or turns the corner; is mainly decorative.

Counterflashing: Flashing set into brick, usually chimneys, covering step flashing.

Cove. Concave wood molding used in interior corners, usually where wall and ceiling meet. Plastic or ceramic members, also concave, fit the corner between counter top and backsplash.

Crawl space: Area between first floor and ground, enclosed by a foundation or open. In houses or other structures with neither basement nor slab on grade.

Cricket: A double-sloped roof member installed between a chimney on a sloping roof and the uphill portion of the roof. Also called saddle, and designed to keep rain and snow from building up along the roof upslope from the chimney. A cricket also is sometimes the term for a triangular piece of nominal 2-inch lumber nailed to an unnotched stair carriage onto which treads and risers are nailed.

Cross-cutting: Sawing lumber across the grain. Opposite of ripping.

Dado: A groove cut across a board.

Decay: Rotting of wood due to a fungus. Cut lumber that has a distinct blue stain is partially decayed.

Dimension lumber: Nominally 2 to 5 inches thick and from 2 to 12 inches wide, in 2-inch increments.

Direct nailing: Perpendicular nailing through one member into another, with nailhead exposed. Also called face-nailing; opposite of blind-nailing.

Door: A member designed to close an opening, including everything but windows.

Dormer: A roofed structure covering an opening in a sloping roof, with a vertical wall with one or more windows. Shed dormers have a sloping roof with one dimension, and are designed to add more space under a roof. "A" or eye dormers are designed primarily for light and ventilation.

Double-hung window: Two sashes installed in vertical grooves that bypass each other when raised or lowered.

Downspout: Vertical metal or wooden tube to direct water from gutters to ground. Also called leader.

Dressed lumber: Lumber planed down from its rough size to its dressed size. A rough 2 x 4, for instance, is a full 2 inches thick and 4 inches wide. A dressed 2 x 4, as generally sold, is 1½ inches thick and 3½ inches wide. A 2 x 4 is actually a nominal 2 x 4 because it is undersized.

Drip: Projection at the edge of a roof to allow rainwater to drip over the edge instead of running down the face of an exterior wall. A groove on the underside of a drip cap or window sill to prevent water from following the contour of the wood and dripping down the wall. The drip actually breaks up the flow of water.

Drip cap: A wood molding set on top of window and door casings to divert rainwater.

Drip edge: A metal strip formed to extend the roof line and acting as a drip.

Dry wall: Plasterboard as an interior wall covering. Installed in sheets, its seams are filled with joint compound; to replace plastering, which is done wet. Sometimes refers to wood paneling.

Dry well: A hole in the ground, sometimes filled with stones or gravel, to allow drainage of water, either from gutters or such things as laundry washers, into the ground.

Eave: Edge of roof along lowest part of roof.

Excavation: A hole in the ground, for cellar, foundation, footings, pipes, or cables.

Expansion joint: An asphalt-impregnated fiber strip placed in grooves in concrete, to prevent concrete from cracking.

Exterior finish: Finish material (siding) such as shingles, clapboards, and vertical siding.

Facia: See Cornice.

Felt: Paper saturated with asphalt, sometimes called tarpaper. Used under roof shingles, siding, or finish floors.

Fieldstone: Natural stone used in construction of building or foundation walls and retaining wall.

Finish: Final coating to a material such as paint, stain, varnish, sealer, or wax. Finish carpentry is the application of exterior and interior trim, sometimes including siding. Finish material usually can be seen. Opposite of rough carpentry.

Fireplace: System to allow building of a fire in an interior room, including firebox, floor, chimney, flue, and mantel.

Fire stop: Cross-blocking of 2 x 4s between studs, midway in height of studs, designed to prevent or retard spread of fire in a wall cavity. Still required in many building codes, but less necessary if cavity is filled with fire-resistant insulation.

Fishplate: A board or plywood connecting beams or rafters butting end to end.

Flagstone: Flat stones, usually cut, used for floors, mostly outside, and for sidewalks, sometimes for retaining walls.

Flashing: Metal placed where roof meets wall or masonry, and in roof valleys, to weatherproof the joint.

Flue: Passage in a chimney for fumes and smoke. Flue lining is clay tubing made in short lengths, and metal in long lengths, to fit inside a chimney. The lining itself makes up a flue. Sometimes old chimneys have no flue lining at all, with the flue being just the passage in the chimney. The latter kinds of chimney are not allowed in most codes.

Fly rafter: End rafter of a gabled roof, overhanging the main wall and supported by lookouts and roof sheathing.

Footing: Concrete platform, wider than the foundation, on which the foundation sits. Installed below the frost line to prevent heaving due to freezing and thawing. Can also support a concrete pier or other types of pillars.

Foundation: Wall, usually of concrete or concrete blocks, that sits on the footing and supports the wooden members of the first floor.

Forms: Wooden members, made of plywood and 2 x 4s, used as retainers for concrete before it sets. Removed after concrete sets. Can be reused.

Framing, balloon: A system of wood framing for a house in which wall studs extend from the foundation sill to the roof line, on which floor joists are attached. Called balloon because of the remark of a builder, when it was introduced in the nineteenth century, that it was no stronger than a balloon. The builder was proved wrong, but balloon framing is not common in today's construction.

Framing, platform: System of wood framing in which each floor is built separately as a platform for the walls. Most common construction of wood-framed houses today.

Frost line: Depth to which ground freezes in winter. Ranges from nothing in the deep South to 4 or 5 feet and more in extreme northern areas of the continent. Footings must be placed below the frost line to prevent heaving or movement due to freezing and thawing.

Furring: Strips of wood, usually 1 x 2 or 1 x 3, or metal, used to even out a rough wall and as a base for securing a finish wall such as plasterboard, plywood paneling, or boards.

Gable: Roof line at the end of a double-sloped roof, forming a triangle from the peak of the roof to the bottom of each end of the rafters.

Girder: Heavy beam of wood or steel to support floor joists. Generally set into the sill and supported at intermediate points by columns.

Glazing compound: A modern putty used to waterproof panes of glass in a wood frame.

Grade: Surface of the ground.

Grain: Direction of fibers in wood.

Groove: A notch running the length of a board, on the edge.

Ground cover: Sheet or roll material (plastic or asphalt felt) used to cover ground in crawl spaces or cellars, to prevent moisture from rising from the ground.

Grounds: Wood strips, ¾ x ⅞ inch, used around wall openings and along floor lines as a surface guide for plastering. Also used for inside corner of an exterior wall against which siding butts.

Grout: Mortar, with or without sand, used to fill joints between floor or wall tile.

Gusset: See Fishplate.

Gutter. Channel of wood or metal used to drain rainwater off a roof and into downspouts. Also called eavestrough.

Gypsum board: See Plasterboard.

Gypsum plaster: A type of plaster, used with sand and water as a base coat. Also used without sand as a top coat.

Hardboard: Manufactured material, made from wood and having wood characteristics, in 4 x 8–foot sheets and thicknesses of ⅛ and ¼ inch. Used as an underlayment for resilient tile.

Header: A beam placed at right angles to floor joists to form openings for chimney, stairways, fireplaces, etc. A beam placed as a lintel over door and window openings.

Header joist: A floor joist connecting the ends of regular floor joists and forming part of the perimeter of the floor framing. Opposite of stringer joist.

Hearth: Floor of a fireplace, inside and outside the firebox, made of brick, stone, or fire brick.

Hip. The sloping ridge formed when two sloping sides of a roof meet. Opposite of valley.

Hip roof: a roof that slopes up from all four sides of a house, meeting at a point in the center of a short ridge. There are no gables in a hip roof.

Humidifier: A device, either appliance-sized and placed in a room or installed in the central heating system, to release water vapor in the house. Used for the health of the occupants and to prevent wood furniture and floors from separating due to shrinkage from loss of moisture.

I-beam: A steel beam, named for its profile shape, used to support joists in long spans, and an as extra-long header over windows or doors.

Insulation: Thermal insulation is placed in wall cavities, in attic floor spaces, and sometimes in cellar ceilings and between roof rafters to reduce escape of heat. It can be fiber glass, rigid or flexible, mineral wool, urethane or styrene, or any other kind of material that reduces heat loss. Sound insulation is of a similar material, mainly fiber glass, and is designed to reduce transmission of sound through walls, ceilings and floors. Reflective insulation is usually aluminum foil in sheet form, designed to reflect heat back into a room, and to reflect outside heat in hot weather. It is ineffective unless an air space is provided between it and the interior wall. If it is used at all, it should be used with thermal insulation.

Interior finish: Material covering interior wall frames, including plaster, plasterboard, or wood.

Jamb: Side and top frame of a window or door.

Joint: Any space between two components.

Joint compound: A plaster type material, containing a glue, used to cover nailheads and joints in plasterboard wall construction. The joints are also covered with paper tape.

Joist: A floor or ceiling beam a nominal thickness of 2 inches, and a depth of 8, 10, or 12 inches, used in parallel to support a floor or ceiling. Floor joists set on the sill and on girders; if there is a second floor, they set on top plates of walls. Ceiling joists set on top plates, and there is no floor secured to them.

Joist hanger: Metal fastener used to secure the end of a joist directly against the side of a girder or other joist. Also called timber support.

Landing: A platform dividing two sections of a flight of stairs, either in a straight stairway or when the stairway makes a right angle turn.

Lap joint: A joint in which one member of a doubled beam or plate overlaps the other member. Most common in wall top plates, made up of 2 x 4s, with a lap joint at each corner.

Lath: Material used as a plaster base: wood, metal, or plasterboard.

Lattice: A framework of crossed wood or metal strips. A board, usually ¼ inch thick and 1½ to 2 inches wide.

Ledger: Strip of lumber nailed to a girder or joist onto which other joists are set. Also a heavy strip nailed to a wall as a joist support.

Level: Horizontal, perpendicular to vertical.

Light: One pane of glass in a window, named for its ability to admit light.

Lintel: Horizontal member supporting the opening above a door or window.

Lookout: In a roof overhang, a short horizontal bracket connecting rafter end with the wall, covered by facia board and soffit. Lookout joists are horizontal members overhanging the main walls and cantilevered over that wall, and are usually employed in bay window construction or when the second floor of a house overhangs the first-floor wall.

Louver: An opening, usually screened, and fitted with angled slats to allow air to move in and out and weather to stay out. Used to ventilate attics and crawl spaces.

Lumber: See Boards, Dimension Lumber, and Matched Lumber.

Mantel: Shelf above a fireplace, including the wood trim around the fireplace opening.

Masonry: Anything like brick, concrete blocks, tile, or other similar material put together with mortar.

Matched Lumber: Boards with tongue in one edge and a groove in the other. Also called tongued and grooved.

Millwork: Lumber shaped or molded in a millwork plant. Usually trim and components of doors and windows.

Miter joint: A joint made by beveling the ends of the pieces to be joined, usually at a 45-degree angle, to form a 90-degree corner.

Molding: Decorative strips of wood, used in trim and detail work on interiors and exteriors of houses.

Mortar: Material used to hold masonry together, made with Portland cement, sand, and lime.

Mortise: A slot or hole cut into wood to receive the tenon of another piece. The mortise is the female portion of a mortise and tenon joint.

Mullion: Vertical divider between two window and (or) door openings.

Muntin: Parts of a window sash frame dividing lights of glass.

Newel, or newel post: Any post to which a railing or balustrade is attached.

Nominal: See Dressed Lumber.

Nosing: Any projecting edge of molding, particularly the projecting part of a tread over a riser, in stairs.

On center: Spacing of studs and joists, measured from the center of one stud or joist to the center of its neighbor. Standard on center is 16 inches; sometimes 24 inches. Abbreviation: o.c.

Panel: A thin piece of wood with slightly thinner edges that are fitted into grooves in stiles and rails of a door. Also applies to this type of construction of finished panel walls. Loosely (and usually called paneling), any wood wall covering, from boards to 4 x 8–foot sheets.

Paper: Term for papers and felts, sometimes impregnated with asphalt and sometimes not, for applying under finish floors, siding, and roofing. Often called sheathing paper and building paper. *See* Felt.

Particleboard: A sheet, in various thicknesses, made by gluing wood chips or particles together. Used as underlayment for resilient tiles and carpeting.

Parting bead: A thin piece of wood, sometimes called parting stop or strip, inserted in the jamb of a window and separating the top and bottom sash. In modern windows, an aluminum or vinyl molding has taken its place and is an integral part of the jamb and the grooves in which the windows ride.

Partition: A wall that subdivides space in a building, usually into rooms.

Penny: Measurement of nails, originally English, indicated price per 100. Abbreviated d.

Pier: A column of masonry, usually heavy, to support structures.

Pigment: Opaque coloring in paint or stain.

Pitch: The slope of a roof.

Plan: Drawing of a building as seen from above, with roof off.

Plaster: A combination of lime and sand and binder, such as gypsum, troweled onto lath, in two or more layers, as a finish wall.

Plasterboard: See Gypsum Board.

Plastic: Any of the modern materials such as urethane, polystyrene, polyethylene, polyvinyl chloride (PVC), and vinyl, used for numerous building purposes.

Plate, sole or floor: Bottom horizontal member of a stud wall, sitting on the subfloor. Top plate: top horizontal member, doubled, of a stud wall, supporting second-floor joists or roof rafters.

Plow, or plough: A groove along the face or edge of a board. The action of making a plow.

Plumb: Vertical; perpendicular to level.

Plywood: Wood made by laminating thin sheets or plies together with grain running in opposite directions. Comes in various sizes and thicknesses, and is particularly useful because it will not split as ordinary lumber boards can. Very strong for its weight.

Post and beam: Type of construction using heavy beams for floors and ceiling, heavier than nominal 2-inch joists and spaced on centers greater than 16 inches. The walls have heavy posts instead of 2 x 4 studs, also set on centers greater than 16 inches. Requires heavy floor, roof, and wall coverings. Utilizes exposed beams extensively. It is a variation of the post and beam technique used up to about 1850.

Preservative: A liquid, with a copper or pentachlorophenol base, used to prevent or retard rot.

Primer: First coat in interior and exterior paint jobs of more than one coat.

Putty: A powdered material that is mixed with water, used to fill nailheads and wood cracks. An obsolete word for glazing compound.

Quarter round: A molding one-quarter of a dowel.

Rabbet: A groove at the end of a board, going across the grain.

Rafter: A beam, nominally 2 inches thick, supporting the roof. A hip rafter forms a hip of a roof (*See* Hip); a jack rafter is a short rafter connecting a hip rafter with the wall top plate, or a valley rafter with the ridge board; a valley rafter forms the valley of a roof, and usually is doubled (*See* Valley).

Rail: Horizontal frame member of a window or paneled door. Also, the upper and lower horizontal members of a balustrade.

Rake: Trim board running parallel to the roof slope to finish off the edge of the roof.

Reinforcing: Steel rods or mesh placed in concrete to strenghthen it. Also called rebars.

Ridge, or ridgeboard, or ridgepole: Horizontal member, nominally 1 or 2 inches thick, forming the ridge of a roof, where the sloping roof surfaces meet at their highest plane.

Ripping: Sawing a board with the grain.

Rise: In stairs, the vertical height of a flight of stairs or the height of one step. In roofs, the vertical height of a roof, from wall top plate vertically to ridge.

Riser: Board enclosing space between treads of a stairway.

Roof: Sloped or flat surface covering the top of a house.

Roofing: Any material on the roof designed to keep out the weather: shingles (asphalt, wood, slate, etc.), metal, roll roofing (asphalt felt and fine gravel made in rolls 3 feet wide).

Rough: Bare framing with wood members. Always covered by a finish of some kind: interior walls, siding, roofing, and trim.

Rough opening: Opening in frame wall for door or window, and in floor for stairwell or chimney.

Run: In stairs, the horizontal length of a stairway. In roofs, the horizontal or level distance over which one rafter runs; half the span of a double-sloped roof.

Saddle: See Cricket.

Sash: A single window frame containing one or more lights of glass.

Sash balance: A spring or weight designed to hold a sash in any position.

Screed: A board used to level fresh concrete. Also a board, usually the thickness of a plaster coat, to act as a guide for the plasterer.

Sealer: A liquid designed to seal the surface of wood as a base for paint, varnish, more sealer, or wax.

Shake: A thick (split, not sawn) wood shingle, used for rustic siding and normal wood roofing.

Sheathing: Exterior covering of a wall. Used as a base for siding. Sometimes the sheathing and siding are combined, such as plywood grooved to look like board on board or reverse board and batten.

Sheet metal work: Nearly everything made of sheet metal, such as gutters and downspouts and warm-air ducts.

Shims: Tapered pieces of wood, generally shingles, used to close gaps between horizontal and vertical wood spaces, usually along floors and between rough openings for windows and doors and the finish jamb.

Shingles: Siding shingles are wood members sawn to a taper, made generally from red or white cedar and tapered from ¼ inch at the butt (bottom) to ⅓₂ inch at the top. Roofing shingles are made of asphalt, metal, slate, etc. Both types are manufactured to standard sizes.

Shiplap: A groove along the side of a board, to allow each board to overlap the other but with their surfaces remaining on the same plane.

Shutter: Hinged exterior covering for a window, usually folded back against the wall; today, nearly 100 percent decorative, and hinged, nailed, or screwed on each side of a window.

Siding: Exterior covering of a wall to keep the weather out and to look good. *See* Clapboard, Shingle, and Board and Batten.

Sill: Sometimes called sill plate: timber sitting directly on masonry foundation; support for floor joists. In windows, the slanting bottom piece of a window frame, designed to shed water.

Sill sealer: Semirigid fiber-glass strip inserted between foundation and wood sill to seal any variations in the foundation or any openings between foundation and sill.

Sleeper: A board, nominally 2 inches thick, secured to a concrete floor to act as a base for a wood floor.

Soffit: The underside of a cornice or boxed eave.

Soil stack: Vent pipe for plumbing, and main plumbing drain for the house plumbing. The same pipe serves both functions.

Span: Distance between supporting points. In a roof, the total level distance between rafter supports.

Square: One hundred square feet, or an area 10 x 10 feet, a unit of measurement of roofing, and sometimes siding. The 100 square feet is the amount exposed to the weather, not the area of the material itself.

Stair: Steps leading up and down.

Stair carriage: Supporting beam for stair treads, a nominal 2-inch beam notched for treads and risers. Sometimes called a stringer.

Stile: Vertical piece of a door or window frame.

Stool: Flat interior molding fitted over window sill, erroneously called a sill.

Stool cap: Apron or support for the stool.

Storm sash: Insulating windows made of wood or aluminum and fitted over the outside of house windows. Sometimes a part of the regular window sash, applied inside or outside.

Story: Living area between floor and ceiling.

Stringer: Support for cross members of openings in a floor. Parallel to joists. Support for stair treads. A stringer joist is the border joist of a floor frame, parallel to intermediate joists. Opposite to header joist.

Strip flooring: Narrow wood floorboards. Plank flooring is wide wood floorboards.

Stucco: Siding made of a plaster made with Portland cement as its base. Applied over metal lath.

Stud: Vertical member in a frame wall. Usually made of 2 x 4 boards. Sometimes made of metal.

Subfloor: Rough boards or plywood secured to floor joists, onto which a finish floor or underlayment is secured.

Suspended ceiling: A ceiling hung from joists by brackets or wires; one not secured directly to joists or furring strips.

Threshold: Wood or metal member usually tapered on both sides and used between door bottom and doorsill. Also used between jambs of interior doors; not generally used in new housing. Sometimes the threshold is an integral part of the sill.

Toenailing: Nailing at an angle, connecting one member with another piece perpendicular to it. Opposite of face-nailing.

Tongued and grooved: See Matched lumber.

Tread: Horizontal board in a stairway that is the part of the step that is stepped on.

Trim: Finish material on interior and exterior of a house, but not including interior walls and exterior siding. Also called woodwork.

Truss: A set of rafters, with collar beam and other connecting members prebuilt and ready to install, connecting opposite wall points.

Undercoat: A primer or sealer for enamel.

Underlayment: A smooth material, plywood, particleboard, or hardboard, installed over a subfloor as a base for finish material such as sheet flooring or tiles.

Valley: Angle formed when two sloping sides of a roof meet.

Vapor barrier: Material, aluminum foil, kraft paper, or polyethylene, designed to prevent passage of water vapor through or into exterior walls. Always placed toward the heated part of the house. Insulation sometimes is made with a vapor barrier. If not, the barrier is secured after insulation is installed.

Veneer: Thin sheets of wood, usually hardwood, applied to core lumber for a fine finish on paneling or furniture. Veneer is made by peeling a log instead of sawing it.

Vent: Pipe, duct, or screened opening to allow passage of air or fumes. Can be used as inlet or outlet.

Wane: Bark or lack of wood, on the edge or corner of a piece of wood.

Weatherstrip: Any material placed at windows and door seams to prevent passage of air. Usually made of wood or aluminum with a vinyl seal.

Welcome: What you'll say when you invite friends in for the housewarming.

appendix 2

Read All About It

Other Publications

While this book is intended to be as comprehensive as possible, no single publication is going to hit all bases, or more appropriately, hit every nail on the head. So, the following list of books and pamphlets is designed for the avid planner of his house who wants as much information as possible.

Some of the techniques in such books are repeated, but there is bound to be more or different information on the less-common techniques used in building a house. Books are listed in alphabetical order.

Build Your Own Home: Guide to Subcontracting, by S. James Hanenau, Holland House Press, 6215 Six Mile Road, Northville, Michigan.

Do-It-Yourself Contracting to Build Your Own Home, by Richard J. Stillman, Chilton Book Company, Radnor, Pennsylvania.

Enjoy and Build It Yourself: Manage the Building of Your Own Home, by George Boggs Roscoe, Acropolis Books, Ltd., Washington, D.C.

Four "Homeowner" handbooks, published by Bounty Books, a division of Crown Publishers, 419 Park Avenue South, New York, N.Y. 10016.

Carpentry and Woodworking, by Richard Day.

Concrete and Masonry, by Richard Day.

Plumbing and Heating, by Richard Day.

Electrical Repairs, by Robert Hertzberg.

The House Building Book: A Professional Builder Demonstrates How He Builds an Entire House, by Dan Brown, McGraw-Hill Book Company , New York, San Francisco, St. Louis and Toronto.

How We Built Our House for Under $10,000, by Walter Rea, Yankee, Inc., Dublin, N.H.

Illustrated Housebuilding, by Graham Blackburn, The Overlook Press, Woodstock, New York, distributed by Viking.

"Wood-Frame House Construction," by L. O. Anderson, Agriculture Handbook #73, Superintendent of Documents, U.S. Government Printing Office, Washington, D.C. 20402, $2. Also available at government book stores in major cities.

"How to Build a Wood-Frame House," a republication of the above Agriculture Handbook #73, Dover Publications, Inc., New York, N.Y.

Thirty-two short but comprehensive circulars published by the University of Illinois Small Homes Council–Building Research Council, are 15 cents each or $4.50 for all 32 (punched for 3-ring binder), also available in a paperbound book for $5, which includes a list of publications. Small Homes Council–Building Research Council, University of Illinois at Urbana-Champaign, 1 East Saint Mary's Road, Champaign, Illinois 61820.

Some of the titles: "Financing the Home," "Maintaining the Home," "Business Dealings with the Architect and the Contractor," "Fundamentals of Land Design," "Hazard-Free Houses for All," "Split-Level Houses," "Household Storage Units," "Kitchen Planning Standards," "Laundry Areas," "Garages and Carports," "Selecting Lumber," "Plywood," "Basements," "Termite Control," "Wood Framing," "Crawl-Space Houses," "Flooring Materials," "Insulation in the Home," "Moisture Condensation," "Chimneys and Fireplaces," "Counter Surfaces," "Window Planning Principles," "Selecting Windows," "Roofing Materials," "Brick and Concrete Masonry," "Heating the Home," "Fuels and Burners," "Electrical Wiring," "Plumbing," "Cooling the Home," and "Interior Design."

The following books are not exclusively about new house building, but present ideas and techniques on specific subjects, which are applicable in new as well as existing houses:

The Family Handyman "Home Improvement Book," by the editors of Family Handyman Magazine, Charles Scribner's Sons, New York, New York.

How You Can Soundproof Your Home, by Paul Jensen and Glenn Sweitzer, Lexington Publishing Company 98 Emerson Gardens, Lexington, Massachusetts.

Manual of Home Repairs, Remodeling and Maintenance, (much of the material reprinted from issues of *Mechanix Illustrated*), Fawcett Books, Grosset & Dunlap, New York, New York.

The Wall Book, Stanley Schuler, M. Evans & Company, New York, New York.

appendix 3

Where in the World?

Sources of Information and Materials

One of the most frustrating experiences for an amateur is to read about a marvelous new product or publication and then not be able to locate it because there was no source listed. This list of sources of information is designed to help cure that problem.

In addition to books listed in "Read All About It," there is a gold mine of information in the form of pamphlets, plans, and other publications available from institutes and manufacturers of every building material. Institutes or associations are organizations representing one aspect of the industry: manufacturers.

It is best to "buy local," that is, buy from local dealers who can be found in the Yellow Pages (Chapter 5). Some of the information on "How to do it" from the institutes and manufacturers is excellent. Some is unabashedly commercial, but that is something that you have to expect. Some of the pamphlets are free, others cost a nominal amount.

This list is not all-inclusive or complete, and listing of manufacturers does not constitute endorsement of their product.

Acoustical Products

Armstrong Cork Co., Lancaster, Pa. 17540.

Celotex Corp., 1500 North Dale Mabry, Tampa, Fla. 33607.

Certain-Teed Products Co., Valley Forge, Pa. 19481.

Dow Chemical Co., Construction Materials, Midland, Mich. 48640.

Johns-Manville, 22 East 40th St., New York, N.Y. 10016.

National Gypsum Co., Buffalo, N.Y. 14217.

Owens-Corning Fiberglas Corp., Building Products Div., Fiberglas Tower, Toledo, Ohio 43659.

PPG Industries, 1 Gateway Center, Pittsburgh, Pa. 15222.

U.S. Gypsum Co., 101 South Wacker Dr., Chicago, Ill. 60606.

Air Conditioning and Heating

Air Conditioning Institute, 1815 Fort Meyer Dr., Arlington, Va. 22209.
Bryant Air Conditioning Co., 7310 Morris St., Indianapolis, Ind. 46231.
Dunham-Bush, Harrisburg, Pa. 22801.
General Electric (heat pumps), Central Air Conditioning Products Dept., Appliance Park, Louisville, Ky. 40225.
Hydrotherm, Inc., Northvale, N.H. 07647.
Westinghouse, Central Residential Air Conditioning Div., Norman, Okla. 70369.
York Co., P.O. Box 1592, York, Pa. 17405.

Aluminum Products

Alcoa Building Products, Inc., Dept. S-73, Suite 1200, 2 Allegheny Center, Pittsburgh, Pa. 15212.
Reynolds Metals Co., Architectural and Building Products Div., 325 West Touhy Ave., Park Ridge, Ill. 60028.

Appliances

Association of Home Appliance Manufacturers, 20 North Wacker Dr., Chicago, Ill. 60606.
Ronson Foodomatic, One Ronson Rd., Woodridge, N.J. 07095.
Sears, Roebuck & Co., Philadelphia, Pa. 19133.

Artificial Glass

American Cyanamid Co., (Acrylic), Plastics Div., Wallingford, Conn. 06492.
Rohm & Haas (Plexiglas), Philadelphia, Pa. 19105.

Bricks, Concrete Blocks, and Masonry

American Precast (concrete stairs and other forms), Building Div., 1468 Main St., Millis, Mass. 02054.

Brick & Tile Service, Inc., Greensboro, N.C. 27405.
Brick Institute of America, 1750 Old Meadow Rd., McLean, Va. 22101.
Duracrete Block Co., 1359 Hooksett Rd., Manchester, N.H. 03104.

BUILDING PRODUCTS

E. I. DuPont de Nemours & Co., Building Products Div., Wilmington, Del. 19898.
Masonite Corp. (hardboard), 29 North Wacker Dr., Chicago, Ill. 60606.
Owens-Corning Fiberglas Corp., Home Building Products Div., Fiberglas Tower, Toledo, Ohio 43659.
Sears, Roebuck & Co., Philadelphia, Pa. 19133.

BUILT-INS (*See* Wood Products)

BURGLAR ALARMS (*See* Security)

CABINETS

Formica Corp., Cincinnati, Ohio 45202.
Georgia Pacific Corp., Georgia Pacific Building, Portland, Ore. 97204.
Long-Bell Div., International Paper Co., Longview, Wash. 98632.
National Kitchen Cabinet Assn., 334 East Broadway, Louisville, Ky. 40202.
Sears, Roebuck & Co., Philadelphia, Pa. 19133.
Triangle Pacific (Gregg & Son, Inc.), 9 Park Pl., Great Neck, N.Y. 11021.
U.S. Plywood, 777 Third Ave., New York, N.Y. 10017.

CEILINGS

Armstrong Cork Co., Consumer Services, Lancaster, Pa. 17604.
Boise Cascade, Wood Products Div., P.O. Box 4463, Portland, Ore. 97208.
Celotex Corp., 1500 North Dale Mabry, Tampa, Fla. 33607.
Conwed Corp., 332 Minnesota St., St. Paul, Minn. 55101.

CERAMIC TILE

Brick & Tile Service, Inc., Greensboro, N.C. 27405.
National Terrazzo and Mosaics Assn., 716 Church St., Alexandria, Va. 23314.
Sears, Roebuck & Co., Philadelphia, Pa. 19133.
Tile Council of America, P.O. Box 326, Princeton, N.J. 08540.

CONCRETE AND CONCRETE FORMING

American Plywood Assn., 1119 A St., Tacoma, Wash. 98401.
Rocktite, Hartline Products, Inc., 2186 Noble Rd., Cleveland, Ohio 44112.

DECKING (*See* Wood Products)

DOORS (*See also* Wood Products)

Morgan Co., Oshkosh, Wis. 54901.
Pease Ever-Strait, Ever-Strait Div., 7100 Dixie Highway, Fairfield, Ohio 45014.
Simpson Timber Co., 2000 Washington Building, Seattle, Wash. 98101.

ELECTRICITY

Edison Electric Institute, 750 Third Ave., New York, N.Y. 10017.

FENCING (*See* Wood Products)

FIXTURES AND ACCESSORIES

Bradley Corp., (faucets), P.O. Box 348, Menomonee Falls, Wis. 53051.
Eljer Plumbingware Div., Wallace-Murray Corp., 3 Gateway Center, Pittsburgh, Pa. 15222.
Miami-Carey (bathroom fixtures), 203 Garver Rd., Monroe, Ohio 45050.
Moen Plumbing Accessories, Zeigler-Harris Corp., 377 Woodland Ave., Elyria, Ohio 44035.

Sears, Roebuck & Co., Philadelphia, Pa. 19133.
Universal-Rundle Corp. (fiber-glass fixtures), New Castle, Pa. 16103.

FIRE PROTECTION (See Security)

FIREPLACES

Vega Industries (heatilator), Syracuse, N.Y. 13205.

FLOORING (*See also* Wood Products)

Amtico (resilient tile), Flooring Div., American Biltrite, Trenton, N.J. 08607.
Armstrong Cork Co. (resilient tile and sheet goods), Consumer Services, Lancaster, Pa. 17604.
Azrock Corp. (resilient tile), 400 Frost Building, San Antonio, Tex. 78292.
Congoleum Industries (resilient tile and sheet goods), 195 Belgrove Dr., Kearney, N.J. 07032.
Flintkote Co. (resilient tile), 480 Central Ave., East Rutherford, N.J. 07073.
Kentile Floors, Inc. (resilient tile), 58 Second Ave., Brooklyn, N.Y. 11215.
Sears, Roebuck & Co. (resilient tile and sheet goods), Philadelphia, Pa. 19133.

FRAMING (*See* Wood Products)

GLASS (*See* Windows)

GRASS

The Lawn Institute, Rte. 4, Marysville, Ohio 43040.

HARDWARE AND LOCKS

Kirsch (drapery hardware), Sturgis, Mich. 49091.
Sargent & Co., 100 Sargent Dr., New Haven, Conn. 06509.
Schlage Lock Co., 2201 Bay Shore Blvd., P.O. Box 3324, San Francisco, Calif. 94119.

Sears, Roebuck & Co., Philadelphia, Pa. 19133.
Stanley Hardware, New Britain, Conn. 06050.

Hatchways

Bilco, New Haven, Conn. 06505.
The Gordon Corp., 504 Main St., Farmington, Conn. 06032.

House Plans

Associated Newspapers, % United Feature Syndicate, 220 East 42nd St., New York, N.Y. 10017.

Humidification

Research Products Corp., Madison, Wis. 53701.

Insulation

Celotex Corp., 1500 North Dale Mabry, Tampa, Fla. 33607.
Certain-Teed Products Co., Valley Forge, Pa. 19481.
Johns-Manville, 22 East 40th St., New York, N.Y. 10016.
Owens-Corning Fiberglas Corp., Home Building Products Div., Fiberglas Tower, Toledo, Ohio 43659.

Intercoms

NuTone Housing Products, Madison and Red Bank Rds., Cincinnati, Ohio 45227.

Lightning Protection

Lightning Protection Institute, 2 North Riverside Plaza, Chicago, Ill. 60606.

Locks (*See* Hardware and Locks)

Molding (*See* Wood Products)

Paint, Varnish, and Stains

Benjamin Moore & Co., Montvale, N.J. 07645.
Darworth, Inc., Avon, Conn. 06001.
Glidden Coatings and Resins, Cleveland, Ohio 44115.
Olympic Stains, 1148 Northwest Leary Way, Seattle, Wash. 98107.
Pierce & Stevens Chemical Corp., 710 Ohio St., Box 1092, Buffalo, N.Y. 14240.
Pittsburgh Paints, PPG Industries, 1 Gateway Center, Pittsburgh, Pa. 15222.
Samuel Cabot, Inc., 1 Union St., Dept. 1975, Boston, Mass. 02108.
Sears, Roebuck & Co., Philadelphia, Pa. 19133.

Paneling (*See also* Wood Products)

Barclay Plastic Finish Wall Paneling, Barclay Industries, Inc., 65 Industrial Rd., Lodi, N.J. 07644.
Boise Cascade, Wood Products Div., P.O. Box 4463, Portland, Ore. 97208.
Masonite Corp., 29 North Wacker Dr., Chicago, Ill. 60606.
Simpson Timber Co., 2000 Washington Building, Seattle, Wash. 98101.
Townsend Paneling (solid hardwood), Potlach Forests, Inc., P.O. Box 916, Stuttgart, Ark. 72160.
U.S. Plywood, 777 Third Ave., New York, N.Y. 10017.
Weyerhaeuser Co., Box B, Tacoma, Wash. 98401.

Patio Structures

Erecto-Pat, 32295 Stephenson, Madison Heights, Mich. 48071.

Plasterboard

GAF Corp., Building Products Div., 140 West 51st St., New York, N.Y. 10020.

Gypsum Assn., 201 North Wells St., Chicago, Ill. 60606.
National Gypsum Co., Buffalo, N.Y. 14217.
U.S. Gypsum Co., 101 South Wacker Dr., Chicago, Ill. 60606.

Plumbing

American Standard, P.O. Box 2003, New Brunswick, N.J. 08903.
Sears, Roebuck & Co., Philadelphia, Pa. 19133.

Plywood (*See also* Wood Products)

Boise Cascade, Wood Products Div., P.O. Box 4463, Portland, Ore. 97208.
Simpson Timber Co., 2000 Washington Building, Seattle, Wash. 98101.
U.S. Plywood, 777 Third Ave., New York, N.Y. 10017.
Weyerhaeuser Co., Box B, Tacoma, Wash. 98401.

Projects and Plans

Craft Patterns Studio, 2200 Dean Street, St. Charles, Ill. 60174. Send for catalogue.
Easi-Bild Pattern Co., Box 215, Briarcliff Manor, N.Y. 10510. Send for catalogue.
Skil Corp., 5033 Elston, Chicago, Ill. 60603.
Stanley Works, New Britain, Conn. 06050.

Publications

Superintendent of Documents, U.S. Government Printing Office, Washington, D.C. 20402. Also at government bookstores in major cities. Send for house/home catalogues.

Roofing

Bird & Son, East Walpole, Mass. 02032.
Certain-Teed Products Co., Valley Forge, Pa. 19481.
Filon Div., Vistron Corp. (translucent roofing panels), 12333 Van Ness, Hawthorne, Calif. 90250.

GAF Corp., Building Products Div., 140 West 51st St., New York, N.Y. 10020.
Philip Carey Corp., Cincinnati, Ohio 45215.

Rust Prevention

Rust-Oleum Corp., 2301 Oakton St., Evanston, Ill. 60204.

Security

ADT Security Systems (burglar, fire alarms), 155 Sixth Ave., New York, N.Y. 10013.
Dytron, 223 Crescent St., Waltham, Mass. 02154.
National Fire Protection Assn., 470 Atlantic Ave., Boston, Mass. 02210.
Stanley Hardware, New Britain, Conn. 06050.

Siding (*See also* Wood Products)

Bird & Son, East Walpole, Mass. 02032.
GAF Corp., Building Products Div., 140 West 51st St., New York, N.Y. 10020.

Stairs

Darworth, Inc. (spiral), Avon, Conn. 06001.
Duvineg Corp. (spiral), Box 828, Hagerstown, Md. 21740.
Morgan Co. (standard), Oshkosh, Wis. 54901.
Whitten Industries (spiral), U.S. Rte. 7, Burlington, Vt. 05250.

Storage (*See* Wood Products)

Stoves

Portland Stove Foundry, 57 Kennebec St., Portland, Maine 04104.
Washington Stove Works, P.O. Box 687, Everett, Wash, 98201.

UNDERLAYMENT (*See* Wood Products)

VACUUMING SYSTEMS

NuTone Housing Products, Madison and Red Bank Rds., Cincinnati, Ohio 45227.

VENTILATION

Home Ventilating Institute, 230 North Michigan Ave., Chicago, Ill. 60601.

WALLPAPER

Imperial Wallpaper Mill, Inc., 3645 Warrensville Center Rd., Cleveland, Ohio 44122.

WATERPROOFING

Philip Carey Corp., Cincinnati, Ohio 45215.

WINDOWS

Andersen Window Corp., Bayport, Minn. 55003.
Morgan Co., Oshkosh, Wis. 54901.
Rolscreen Co., Pella, Iowa 50219.

WOOD PRESERVATION

American Wood Preservers Institute, 2600 Virginia Ave., N.W., Washington, D.C. 20037.

WOOD PRODUCTS

This category covers everything made of wood or wood by-products. The following organizations are listed with their specialties:

American Wood Council, 1619 Massachusetts Ave., N.W., Washington, D.C. 20036 (has information on paneling, siding, storage, built-ins, underlayment, sheathing, subflooring, roof decking, concrete forms, fencing, decks, framing, flooring, doors, windows, moldings, roofing, interior wall coverings and door and window frames).

The council is made up of the following associations which have information on the following subjects:

Paneling, siding, storage, built-ins, underlayment: American Hardboard Assn., 20 North Wacker Dr., Chicago, Ill. 60606.

Paneling, siding, sheathing, subflooring, underlayment, storage, built-ins, roof decking, concrete forming: American Plywood Assn., 1119 A St., Tacoma, Wash. 98401.

Paneling, siding, fencing, decks: California Redwood Assn., 617 Montgomery St., San Francisco, Calif. 94111.

Framing, sheathing, siding, paneling, fencing, decks: Canadian Wood Council, 300 Commonwealth Building, 701–170 Laurier Ave., West, Ottawa, Ont. K1P 5V5, Canada.

Hardwood flooring: National Oak Flooring Manufacturers Assn., 814 Sterick Building, Memphis, Tenn. 38103.

Underlayment, storage, built-ins: National Particleboard Assn., 2305 Perkins Pl., Silver Spring, Md. 20910.

Doors, windows, moldings: Ponderosa Pine Woodwork, 39 South LaSalle St., Chicago, Ill. 60603.

Roofing, siding, interior wall coverings: Red Cedar Shingle and Handsplit Shake Bureau, 5510 White Building, Seattle, Wash. 98101.

Framing, sheathing, decks, siding, interior wall coverings: Southern Forest Products Assn., P.O. Box 52468, New Orleans, La. 70150.

Siding, interior wall coverings, decks, framing: Western Red Cedar Lumber Assn., 700 Yeon Building, Portland, Ore. 97204.

Wood moldings, door frames: Western Wood Moulding and Millwork Producers, P.O. Box 25278, Portland, Ore. 97225.

Framing, sheathing, siding, interior wall coverings, decks, fencing: Western Wood Products Assn., 700 Yeon Building, Portland, Ore. 97204.

Other firms and their specialties:

Boise Cascade, Wood Products Div. (all products), P.O. Box 4463, Portland, Ore. 97208.

E. L. Bruce Co., Div. of Cork Industries, Inc. (hardwood floors), 1648 Thomas St., Memphis, Tenn. 38101.

Georgia Pacific (all products), 900 S.W. Fifth Ave., Portland, Ore. 97204.

Masonite Corp. (paneling, underlayment, siding), 29 North Wacker Dr., Chicago, Ill. 60606.

Simpson Timber Co. (all products), 2000 Washington Building, Seattle, Wash. 98101.

Townsend Paneling (solid hardwood paneling), Potlatch Forests, Inc., P.O. Box 916, Stuttgart, Ark. 72160.

U.S. Plywood (plywood and plywood products), 777 Third Ave., New York, N.Y. 10017.

Weyerhaeuser Co. (all products), Box B. Tacoma, Wash. 98401.

Index